T0275912

Insect overwintering is a fascinating process involving many physiological, epidemiological, biochemical and behavioural changes. The study of the overwintering process can offer an insight into the development of insects, as well as help us to predict the patterns of disease epidemic and crop destruction caused by some species.

This book provides a comprehensive account of the various forms of insect overwintering that highlights important areas of economic interest. It will be essential reading for advanced students and researchers in the fields of zoology, agriculture, forestry and ecology.

THE ECOLOGY OF INSECT OVERWINTERING

THE ECOLOGY OF
INSECT OVERWINTERING

S. R. LEATHER
Lecturer in Pest Management,
Department of Biology,
Imperial College

K. F. A. WALTERS
Principal Scientific Officer,
Central Science Laboratory, MAFF,
Harpenden

and

J. S. BALE
Professor of Environmental Biology,
Birmingham University

CAMBRIDGE
UNIVERSITY PRESS

CAMBRIDGE UNIVERSITY PRESS
Cambridge, New York, Melbourne, Madrid, Cape Town, Singapore, São Paulo

Cambridge University Press
The Edinburgh Building, Cambridge CB2 2RU, UK

Published in the United States of America by Cambridge University Press, New York

www.cambridge.org
Information on this title: www.cambridge.org/9780521417587

First published 1993
Reprinted 1995
First paperback edition 1995

A catalogue record for this publication is available from the British Library

Library of Congress Cataloguing in Publication data
Leather, S. R. (Simon R.)
The ecology of insect overwintering / Simon R. Leather, Keith F. A.
Walters, and Jeffrey S. Bale.
  p. cm.
Includes bibliographical references (p.) and index.
ISBN 0–521–41758–9 (hc)
1. Insects – Ecology. 2. Insects – Physiology. I. Walters, Keith
F. A. II. Bale, Jeffrey S. III. Title.
QL463.L43 1993
595.7′0543 – dc20 91-43260 CIP

ISBN-13 978-0-521-41758-7 hardback
ISBN-10 0-521-41758-9 hardback

ISBN-13 978-0-521-55670-5 paperback
ISBN-10 0-521-55670-8 paperback

Transferred to digital printing 2006

1

Introduction

Although most insects in temperate climates spend a large proportion of their life in an overwintering stage (the small willow aphid, *Aphis farinosa* (Gmelin) (Homoptera: Aphididae), for example, spends 75 per cent of the year as an egg) the study of insect overwintering has, in many cases, been surprisingly neglected. Perhaps this is a reflection of the fact that many entomologists, like the majority of the insects they study, spend the long cold winter months carefully insulated from the outside world!

Why study overwintering?

The vast majority of the prodigious amount of literature concerning insects, and particularly the literature concerning those insects of economic importance (with which this book is mainly concerned) details investigations of the summer stages of the life cycle. This is, of course, understandable – the insects are present in large numbers during the summer (generally in the multiplicative stage of the life cycle) and this is, in general, the time when the damage to the crop becomes apparent. In addition, the majority of control measures are applied at or just before this time of the year. However, it is a sometimes forgotten fact that the size of the insect populations entering the overwintering stages, and the subsequent survival of these stages, play a major part in determining the population levels encountered in the following spring and summer. Although it has been stated that the literature on overwintering is extensive (Danks 1978), it has too often been of a superficial nature or confined to one specific area, generally that of cold-hardiness. Further, relatively few attempts have been made to correlate the ecological information gained from field studies with the results of physiological and biochemical investigations normally conducted in the laboratory. This is a serious fault and should be remedied.

Recent work within allied groups, e.g. red spider mite, and within insect groups such as the Aphidoidea, has highlighted the advantages, in terms of control and prediction, to be gained from a detailed knowledge of overwintering habits. For example, the number of overwintering eggs of the aphids *Aphis fabae* Scopoli. and *Rhopalosiphum padi* (L.) (Homoptera: Aphididae) can be used to forecast the summer populations of these species developing on field beans in England and on cereals in Finland, respectively (Way *et al.* 1981; Leather 1983). In addition, the spread of potato leaf roll virus can be predicted from a knowledge of the overwintering habits of its aphid vectors (Turl 1983). In the field of forest entomology, the numbers of overwintering pupae of the pine beauty moth, *Panolis flammea* (D & S) (Lepidoptera: Noctuidae), and the pine looper moth, *Bupalus piniaria* (L.) (Lepidoptera: Geometridae), can be used to predict the need for control measures in the coming season (Stoakley 1977; Bevan and Brown 1978). There are many other similar examples, and these will be dealt with in greater detail later (see Chapter 6). There are also many examples where a knowledge of the overwintering biology of an insect would be advantageous, and some of these are pointed out. The study of insect overwintering habits is thus of great importance to entomologists.

What is overwintering?

This may at first seem a relatively simplistic question to pose, and one that may be answered just as simplistically as 'the way that an organism passes the winter'. However, this does not get us very far. Many different definitions of overwintering have been suggested, illustrating the complexity of the subject. Mansingh (1971) considers overwintering in his discussion of dormancy under the heading of hibernation. He defines insect hibernation as 'a physiological condition of growth retardation or arrest, primarily designed to overcome lower than optimum temperatures during winter or summer'. He goes on to point out that overwintering insects also have to contend with the other adversities associated with winter conditions. The main difference between hibernating insects and active ones is that the optimum body temperatures of the former are lower, which leads to torpidity and other metabolic changes. Under the heading of hibernation there are several subclassifications – after all, not all insects respond to winter conditions in the same way – some remain active throughout most of the winter, others cease activity completely, and many species adopt a mixed strategy depending on the daily weather patterns. These subclassifications have been defined by Mansingh as:

1. *Quiescence.* This is the response of individual insects to a sudden unanticipated, non-cyclic and usually short duration deviation of normal weather conditions. This is probably a phenomenon confined to early winter or to winter active insects, and only results in growth retardation.
2. *Oligopause.* This is seen in species inhabiting areas of moderate winter where there is a fixed period of dormancy in response to a cyclic and rather longer term climatic change. These species show a greater and longer retardation in growth than quiescent insects and growth arrest often occurs. Nutritional reserves are required, although periodic feeding does occur during the winter.
3. *Diapause.* This is the most highly evolved system of dormancy for overcoming cyclic, long term extremes in environmental conditions. The main differences between diapause and the two systems described above are that in diapause: (i) there is a definite preparatory phase, usually initiated by a temperature-independent factor, e.g. photoperiod, which involves metabolic changes; (ii) the insect does not feed during the winter; and (iii) the return of favourable conditions does not terminate diapause immediately. Rather, a complex series of events, e.g. the accumulation of heat units or critical photoperiods, is required before the insect terminates diapause and begins to emerge from its overwintering state.

Advantages and disadvantages of overwintering

This must be looked at in two ways – first in comparison to insects in tropical climates and secondly as compared to those insects that, although in temperate or extremely cold climes, remain active. It is important to recognise that temperate and polar species have to overwinter and that any advantages or disadvantages follow from this requirement.

The main advantage of overwintering for a temperate or polar insect, compared to a tropical insect showing a continuous uninterrupted life cycle, is that overwintering in a sufficiently cold-hardy state allows species to survive in inconstant environments that are not favourable for continuous reproduction and normal metabolic functions. A spin-off advantage from this is that predation by winter active animals, e.g. birds, is reduced because the resistant overwintering stages are generally more difficult to locate than the active feeding stages. Overwintering also has an advantage, not often recognised, that in strongly, but not entirely, parthenogenetic species, the production of overwintering eggs provides an opportunity to reproduce sexually without sacrificing too much of the rapid multiplicative phase of their life cycle (Ward *et al.* 1984). In addition, overwintering insects in full diapause are often more resistant to cold than those remaining active during the winter and do not have to face the danger of starvation, which can be a major mortality factor of winter active insects (Mansingh 1971). Some tropical insects respond to adverse conditions, e.g. extreme heat or drought, in a similar fashion, i.e. they aestivate, or go into summer diapause (Tauber *et al.* 1986).

There are, however, certain disadvantages accruing to overwintering insects. Although predation is reduced, once found, they are more vulnerable because they are usually non-mobile. This lack of mobility also means that the insect is unable to move from its overwintering site and so, if conditions become less favourable – drying out or waterlogging of the overwintering site being prime examples – they must be endured. There is thus often a high natural overwintering mortality rate underlying that imposed by predation, e.g. eggs of the bird cherry-oat aphid, *Rhopalosiphum padi*, decline by 80 per cent over the winter (Leather 1981) and overwintering pupae of the pine beauty moth, *Panolis flammea*, habitually suffer up to 40 per cent mortality (Leather 1984).

It can be seen that there are a number of points for and against overwintering and these will be discussed at greater length later.

2

The overwintering locale – suitability and selection

Introduction

Overwintering insects can be exposed to extreme physical conditions and the choice of the site in which they spend the winter is often critical to their survival. By locating a suitable overwintering site before the onset of harsh conditions, insects can mediate the adverse effects of low temperatures, rapidly changing temperatures, the chilling effect of wind, desiccation, lack of oxygen in aquatic environments and other similar hazards.

The differences in suitability, both between and within winter habitats can best be understood by considering those factors affecting the conditions experienced in a particular habitat. Clearly, the general conditions that have to be contended with, depend on the regional climate, which is determined partly by latitude, partly by altitude and partly by proximity to the seas, lakes and mountains. These general conditions are modified by local effects, such as inclination and aspects of slopes, vegetation, the nature of the ground surface and snow cover (Flohn 1969).

Temperature, of both macro- and microhabitats, is a useful factor with which to compare insect overwintering sites (Danks 1978) and will be used to illustrate how both regional climate and local conditions affect the suitability of sites for winter survival of insects.

The winter habitat – regional climate

Climate on a regional scale is partly determined by continental position. Lower winter temperatures generally occur in the centre of large land masses, well away from large bodies of water and less severe winter conditions are found in areas nearer the sea. This is because land surfaces

heat up and cool down more rapidly than the surface of water. The faster
relative rate of heating and cooling of land surfaces also results in a much
greater variability in daytime and night-time temperatures.

This difference in the rates of heating and cooling in the two habitats
is largely due to differences in four physical properties of land and water
(Strahler 1963). Firstly, water has a higher specific heat than soil or rock.
Secondly, because of the mixing of warmer surface layers with cooler
layers at greater depth heat is distributed more quickly in water than on
land. Thirdly, evaporation from water surfaces prevents temperatures
from rising as high as they otherwise would and, finally, as water is
transparent, it allows the transmission of solar radiation to a depth of
several metres. (Solid ground surfaces absorb all the energy in their top
few centimetres.) Thus, the proximity of the overwintering habitat to large
bodies of water determines in part the severity of the conditions experienced
by an insect during the winter.

The effect of position in large land masses can be modified by advection,
a process involving the transfer of heat by the horizontal movement of
air. Thus, a coastal site in which the prevailing winds bring air from a
continental interior will have more of a continental-type temperature cycle
than one where the winds come from the sea. Wind also has a chilling
effect on exposed overwintering sites and is thus a factor to be considered
on both a regional and local scale.

Latitude

Latitude exerts a major influence on both the average temperature
throughout the year and the seasonal extremes of summer and winter.
These differences occur because of variations in the annual cycle of
incoming solar radiation. At the equator there is relatively small variation
in insolation throughout the year. However, at higher latitudes insolation
reaches a peak in midsummer and minimum levels occur in the coldest
month, January. As latitude increases, values for insolation during the
winter period drop very rapidly until, north of the arctic circle, and south
of the antarctic circle there is a total lack of incoming solar radiation for
a period of up to 6 months (Strahler 1963).

Such variations in the severity of winter at different latitudes can
frequently be related to differing adaptation shown by individuals of the
same species. For example, it has been suggested that nymphs and adults
of the cereal aphid, *Sitobion avenae* (Homoptera: Aphididae), survive the
relatively mild winters of the southern part of Britain but that at more

northerly latitudes the species relies to a greater extent on the more cold-resistant egg stage (Walters and Dewar 1986). Responses other than those related to temperatures have also been linked to latitude. The larvae of the pitcher plant mosquito, *Wyeomia smithii* (Diptera: Culicidae), spend the winter in a state of developmental arrest that is evoked, maintained and terminated by photoperiod. The critical photoperiods are strongly related to latitude, this factor accounting for 80.5 per cent of the observed variation (Bradshaw 1976).

Altitude

Altitude also affects the winter conditions experienced by insects. As the greenhouse effect is less pronounced at higher altitudes, high altitude overwintering sites usually have both lower average air temperatures and larger daily temperature ranges than low altitude sites. For example, the critical period for the initiation of developmental arrest in *W. smithii* is affected by altitude (Bradshaw 1976) as well as latitude. To account for geographic differences in the timing of such phenological events as insect emergence, Hopkins (1938) suggested that an increase of one degree of latitude is equivalent to a rise of about 122 m in altitude. This is still a generally accepted principle.

Terrain and overwintering success

Variations in topography can create significant differences in weather conditions in what might at first appear to be homogeneous habitat. For example, deep valleys that are invariably without cloud can lie next to ridges or hill tops that are rarely without cloud. On a smaller scale, a footprint in a lawn can be a frost hollow, whereas the rest of the lawn can be frost-free. Thus, the influence of terrain in the overwintering locale of insects should not be overlooked. The influence of terrain on insect population dynamics is discussed elsewhere (Wellington and Trimble 1984); we will confine our discussion to the influence of terrain on the suitability of overwintering locale and its effect on site selection.

Inversion effects and air mass stagnation

Observations of outbreaks of the moth *Epirrita* (= *Oporinia*) *autumnata* (Lepidoptera: Geometridae) in northern Sweden (Tenow 1975) have shown that some groups of birch trees were unaffected, although trees surrounding them were completely or partially defoliated – the so-called 'green island' effect. This was attributed to the fact that these trees were in cold-air 'pools' or 'lakes' that accumulated during wintertime inversions. The overwintering eggs of *E. autumnata* on trees in these cold air 'lakes' were killed, while trees above the surface of the 'lake' were defoliated to a greater or lesser extent depending on their degree of immersion. After a particularly severe winter, these 'green islands' joined together and reached uphill to the top of the inversion as temperatures below that level had fallen sufficiently to kill all the eggs present. When the supercooling points of the eggs of *E. autumnata* during various stages of winter were compared with the profiles of minimum temperatures experienced, a very good correlation between the horizontal and vertical extent of undamaged forest was found (Tenow and Nilssen 1990).

A similar situation is seen in the black pineleaf scale, *Nuculaspis californica* (Homoptera: Diaspididae), on Douglas fir, *Pseudotsuga menziesii*, in the western United States. In normal circumstances the overwintering scale on the needles is cold conditioned before freezing occurs and it persists without significant winter mortality. However, in some of the valley systems cool air from the higher surrounding areas drains into the lower valleys. If the mouths of the valley are narrow enough to prevent outflow, this cool air accumulates in the valley bottom. When the regional weather is near freezing, temperatures in such valley bottoms soon fall well below freezing point and kill all the scales, only those further up the slope surviving. In some years, if the regional air is not so cold, cold conditioning occurs in the valley bottoms and it is the up-slope scales that suffer the greatest overwintering mortality (Fig. 2.1) (Edmunds 1973).

The lodgepole needle miner, *Coleotechnites starki* (Lepidoptera: Gelechiidae), also shows a similar pattern of overwintering mortality in the Canadian Rocky Mountains. However, in this case the localised freezing conditions are caused by a combination of prolonged air mass stagnation caused by two opposing airflows meeting and continual radiant cooling from the surface layer of the air mass. This results in temperatures in the valley bottoms falling as low as $-50\,°C$ while up-slope temperatures only a few hundred metres away may be only $-10\,°C$. Unless insulated by snow, the needle miners will inevitably die. Only the up-slope individuals near the top of the inversion within the warmer temperature zone have a chance of survival under these conditions (Henson *et al.* 1954; Stark 1959a,b).

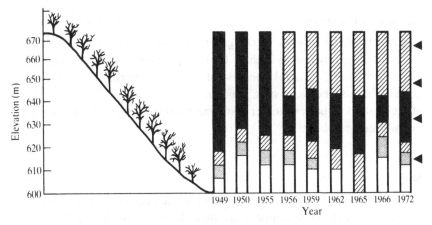

Fig. 2.1. Infestation of the black pine leaf scale, *Nuculaspis californica*, on Douglas fir in Spokane County, Washington, USA, showing the annual variation in population size resulting from the effect of frequent rapid freezing below 610 m, episodic rapid freezing between 610 and 620 m and above 640 m, and annual cold conditioning before freezing between 620 and 640 m. ■ scale severe killing trees, ▨ scale moderate – damage light, ▣ scale rare, □ scale absent, ▶ elevations at which measurements were made (after Edmunds 1973).

Storm fronts and jet streams

Those insects that adapt the foliage of their host plants as shelters against predators and weather during their larval life often use the same shelters as winter hibernacula. The spruce budworm, *Choristoneura fumiferana* (Lepidoptera: Tortricidae), is such an insect and although the greenhouse effect of these shelters is ideal during the growing period of the larvae, they can become detrimental to overwintering pupae in areas where winter sunshine is at a premium. Early snow cover kept in place on the conifer branches by continuing cool weather provides the best overwintering conditions for spruce budworm, but in some regions repeated rapid thawing and freezing of snow can be brought about by the frequent frontal storms. Under such conditions the local terrain can markedly influence the effect of the continental storm track and its associated jet stream, and thus exert a considerable influence on winter survival. Thus, insect populations will often remain small unless a climatic release occurs in a region where suitable forest is available (Wellington *et al.* 1950; Wellington 1952, 1954a,b).

Effects of local habitat

The effects of the severity of the general regional climate during winter can be modified by insects both by the stage in which they overwinter (see Chapter 5) and by careful selection of the overwintering site. There are great differences between sites, which may be above or below the surface of the ground or near or under water, and insects exploit all such sites as overwintering habitats.

Soil

During the winter, soil generally offers both a warmer environment and one with a more stable temperature than the air above and is frequently utilised by overwintering insects. Soil temperatures can be influenced by a number of factors such as depth below the surface, the physical characteristics of the soil, moisture content and the presence and depth of a snow covering, all of which influence insect survival.

Soil depth and soil type

Heat transfer in the soil is by conduction. However, as mineral matter is not a particularly good conductor of heat the effects of surface heating and cooling die out rapidly with depth. Hence, the range of daily temperature variations decrease with increasing depth thus providing a more stable temperature environment during winter. Among the insects that take advantage of this stable environment is the grasshopper, *Melanoplus bivittatus* (Orthoptera: Acrididae), which overwinters in the Canadian prairies. The egg pods of this species are laid in the soil at a depth of approximately 5 cm. The soil temperature at that depth during a normal winter is significantly higher than air temperatures and although the mortality of eggs will increase with a decrease in soil temperatures, the temperatures which cause substantial mortality ($-15\,°C$ or lower) do not usually occur (Mukerji and Braun 1988).

The poor ability of soil to conduct heat can also offer other potential advantages to overwintering insects. During the cold winters of the continental interiors in northerly latitudes, the period of solar radiation is short and there is little warming of the soil surface. As cooling of the surface layers may occur for up to 18 hours in every 24 in such locations, the soil also becomes warmer with depth during the winter period (Fig. 2.2). Many overwintering insects take advantage of the higher temperatures to be found in deeper soil layers. However, because of the increased

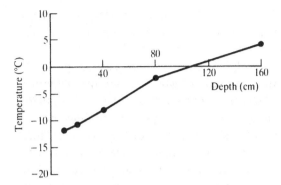

Fig. 2.2. January soil temperatures at depths of 10–160 cm in the USSR (after Strahler 1963).

costs of burrowing the deeper they go, and the fact that improved spring conditions effectively come later to deeper layers, insects cannot simply burrow as deeply as they need to escape the cold (Danks 1978). Instead, they have to balance the advantages of escaping harsh winter conditions with the disadvantages of burrowing well into the soil. Larvae of the broad-necked root borer, *Prionus laticollis* (Coleoptera: Cerambycidae), illustrate this trade-off. These insects are killed if subjected to temperatures of 0 °C or lower, and in the field they migrate downwards to keep just ahead of the advance of subfreezing temperatures (Benham and Farrar 1976).

In some specialised species there can be a disadvantage in digging too far into the surface layers. For example, the parasitoids *Muscidifurax zaraptor*, *Spalangia cameroni* and *Urolepis rufipes* (Hymenoptera: Pteromalidae), which attack house flies, *Musca domestica* (Diptera: Muscidae), and stable flies, *Stomoxys calcitrans* (Diptera: Muscidae), overwinter in dung heaps in eastern Nebraska. Survival was found to be better near the surface as those insects overwintering slightly deeper did not become sufficiently cold acclimated to survive the periodic thawing and refreezing of the dung (Guzman and Petersen 1986).

Temperature differences between soil types in the same geographical area have frequently been reported and are usually due to the differences in heat conductivity in soils with different particle sizes. It was once thought that such differences between soil types were too small to have a significant effect on insect survival in winter (Mail 1930), but Danks (1978) has since pointed out that even small temperature differences may result in important, near lethal temperatures. Only a few examples of soil type influencing winter survival in insects have been reported. The onion maggot, *Delia antiqua* (Diptera: Anthomyiidae), is an important agricultural pest in the United States and can cause losses of up to 40 per cent of cultivated onions (Haynes *et al.* 1980). It overwinters as a pupa in the soil from late October to the beginning of May. Among overwintering pupae subjected to a range of soil types and depths, Whitfield *et al.* (1986) have found that survival ranged from 71 to 83 per cent in different soil types and from 82 to 94 per cent in similar soils at different depths. Thus, although both soil type and depth of overwintering site were shown to have some effect on the survival of the insect, these effects were small. It was concluded that in this case the insects were probably relatively immune to abiotic mortality factors, and the biotic mortality factors (e.g. pathogens or natural enemies) were either lacking or inefficient.

Soil moisture and surface ice

The relationship between soil temperature variations and variations in the temperature of the air above is also influenced by the water content of the soil and the thickness of ice in the surface layers (Holmquist 1931). When the moisture content of deeper layers is maintained at higher levels by frozen surface layers, soil temperature is affected by the buffering effect of the water-saturated habitat, which slows the rate of cooling relative to the air (Baust 1976). Thus, insects that burrow to slightly deeper layers are afforded further protection from both extremely low air temperatures and wide variations in temperatures over short periods of time.

A high soil moisture content maintained by frozen surface layers is also beneficial to those insects that are susceptible to desiccation during winter. Winter survival of corn rootworms, *Diabrotica virgifera* and *D. longicornis* (Coleoptera: Chrysomelidae), for example, is enhanced in moist soil (Calkins and Kirk 1969).

Insects that have not burrowed deeply enough may still be susceptible to death by freezing if more than the superficial surface layers are frozen. The depth of soil frozen depends on aspect (north-facing being more susceptible to freezing than south-facing), period and severity of exposure, and slope of the ground. The freezing of soil can be continuous, as in the Antarctic, or periodic, as in temperate regions, and the overwintering strategy of a particular insect must take all this into account.

Snow and soil temperature

A covering of snow on the soil surface is frequently an important factor in insect survival (Hayhoe and Mukerji 1987). The insulating effect of snow varies with its density but there is a roughly log-linear relationship between ground temperature and snow depth (Mackay and Mackay 1974).

Even relatively shallow snow can significantly enhance the survival of insects overwintering in the soil. Mortality of overwintering larvae and pupae is an important factor in the population dynamics of the bertha armyworm, *Mamestra configurata* (Lepidoptera: Noctuidae), and outbreaks appear to occur in areas where winter soil temperatures are highest. Lamb *et al.* (1985) found that there was a higher survival of pupae in soil covered by deep snow than shallow snow, because of higher soil temperatures in the former case. Pupal survival was higher than 90 per cent under snow depths greater than 20 cm and negligible when there was no snow cover (Turnock *et al.* 1983). The way in which insects utilise the insulating properties of soil and snow is also well illustrated by adults of the rice water weevil, *Lissorhoptrus oryzophilus* (Coleoptera: Curculionidae), which has been found overwintering in shallow soil covered by snow where the minimum temperature reaches about −2 °C, compared to a minimum temperature for the region of −19 °C (Kobayashi *et al.* 1988).

Sites above the surface of soil

Winter is a period that can impose a severe physiological stress on insects, and overwintering sites that are above the soil surface are generally colder and experience more variable temperature regimes than those below the surface. However, winter severity in such sites varies both temporally and spatially and many insects are adapted to take advantage of the milder conditions that can occur.

In the cool temperate climate of countries such as the UK there is sufficient annual variation in both the total number of frost days and the frequency of severe frosts to exert a considerable influence on the population dynamics of insects. Such variation in winter severity is well illustrated by meteorological records from the UK. Between 1975 and 1984 the number of air frost days per winter in northern England ranged from 50 to 100, with a mean of 67 ± 6. At the soil surface (grass minimum temperature) the number of frost days during the same period varied from 100 to 160, with a mean of 140 ± 5. During the winter of 1975–6 the lowest temperature did not fall below −10 °C, whereas in other years temperatures below −15 °C were not uncommon.

The differential occurrence of frosts at the soil surface compared to the air above, which was illustrated by these data from northern England, is

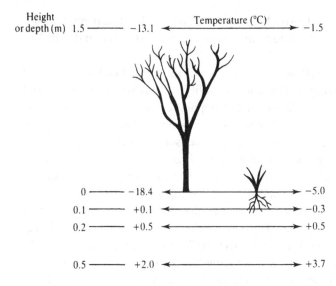

Height
or depth (m) 1.5 ———— −13.1 ←———— Temperature (°C) ————→ −1.5

0 ———— −18.4 ←————————————→ −5.0
0.1 ———— +0.1 ←————————————→ −0.3
0.2 ———— +0.5 ←————————————→ +0.5

0.5 ———— +2.0 ←————————————→ +3.7

Fig. 2.3. Vertical stratfication of temperature in winter in the environs of a tall
hedge in northern England (after Bale *et al.* 1988).

an important factor to be considered in studies of insect overwintering. A
comparison of the temperatures at the soil surface and at the top of a
1.5 m hedge on a very cold night and during an average frost is shown
in Fig. 2.3. Such spatial differences in temperature are important not only
in their effect on insect survival, but also in determining which measurements should be taken by experimenters when comparing mortality
counts with environmental temperatures. For example, a standard air
measurement in a meteorological screen may underestimate the level of
cold experienced by aphids on a crop of winter cereals, which could be
overwintering at 5–10 cm above the soil surface.

The harsher environment offered by habitats exposed to air can be
mediated to a varying degree by some kind of physical protection such
as a covering of snow or ice. Temperatures of overwintering sites above
the snow can be influenced by aspect; warmer conditions occurring on
south-facing sites due to the effect of insolation (Jensen *et al.* 1970) and
cooler conditions resulting from exposure to wind. Indeed insolation even
affects insects that hibernate in somewhat protected conditions, such as
ichneumonid wasps beneath tree bark (Dasch 1971).

Finally, whatever the severity of the conditions they face, insects can
increase their chances of survival by facing the coldest weather in the
stage of their life cycle that is most resistant to low temperatures (see

Chapter 5), by selecting overwintering locations that offer some protection from the rigours of winter or by behaviour that mediates the effects of adverse conditions.

Winter severity and insect survival above the soil surface

The temporal and spatial variations in winter severity outlined in the introduction to this section exert a considerable influence on the population dynamics of some species of insects.

The aphids (Homoptera: Aphididae) offer a range of examples to illustrate this. Species such as the lupin aphid, *Macrosiphum albifrons*, are thought to be sufficiently cold-tolerant to survive as nymphs and adults and reproduce through most British winters (Carter and Nichols 1989). Low winter temperatures have relatively little effect on these aphids.

In Scotland the peach-potato aphid, *Myzus persicae*, the potato aphid, *Macrosiphum euphorbiae*, and the greenhouse and potato aphid, *Aulacorthrum solani*, overwinter as nymphs and adults on wild plants such as small nettle, *Urtica urens*, and groundsel, *Senecio vulgaris*. The success of such overwintering depends upon the relative severity of the winter, with a greater survival rate in milder conditions (Fisken 1959; Turl 1983; Walters 1987). A series of field experiments conducted over a period of 6 years have shown that the temperatures experienced in March (Fig. 2.4) play an important role in determining the size of the populations of *M. persicae* present in spring (Walters 1987).

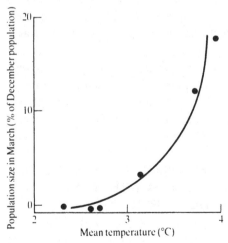

Fig. 2.4. The reduction in the size of overwintering *M. persicae* populations between December and March in relation to mean daily temperatures during that period (after Walters 1987).

The grain aphid, *Sitobion avenae*, also overwinters as nymphs and adults (anholocyclically) in the milder winters of southern England and, once again, the extent to which such colonies survive the winter is related to the severity of the weather (Dewar and Carter 1984). However, in Scotland the harsher winter weather prevents successful anholocyclic overwintering in most years and the species relies to a greater extent on the more resistant egg for winter survival (Walters and Dewar 1986). The closely related aphid, *Sitobion fragariae*, only appears to overwinter successfully under British conditions in the resistant egg stage, irrespective of the latitude of the overwintering site (Hille Ris Lambers 1950; Walters and Dewar 1986).

Protected habitats and choice of overwintering site

Not all sites above soil level are exposed to the major temperature variation of the air. For example, decaying tree stumps tend to dampen variations in ambient temperatures due to the insulative properties of porous wood (Baust 1976). Thus, insects such as the carabid *Pterostichus brevicornis* (Coleoptera: Carabidae), which overwinter in stumps, are not subjected to such rapid cooling rates as they would be if they overwintered in exposed sites.

A wide variety of protected overwintering habitats are exploited by many insect species. The grain aphid, *S. avenae*, which, as described above, overwinters in the southern half of the UK as nymphs and adults – stages that are susceptible to low temperatures – gains some protection by crawling into the middle of dense grass tussocks and hedgerows (Dean 1974). The blackfly, *Simulium pictipes* (Diptera: Simuliidae), spends the winter as eggs laid in crevices in rocks at the edges of streams (Kurtak 1974).

Choice of habitat plays an important role in the survival of the gypsy moth, *Lymantria dispar* (Lepidoptera: Lymantriidae), which overwinters as eggs laid on tree trunks. The critical feature in determining their survival in low winter temperatures is that the eggs are laid low enough to take advantage of the insulating properties of snow cover (Leonard 1972).

Thus, as we have already seen, the choice of the habitat in which the insects overwinter can be critical to their survival. The mechanisms by which sites are selected are discussed later in this chapter.

Thermoregulation

Several insects which overwinter in sites above the soil surface have been reported to thermoregulate either by behaviour that utilises an external source of heat (May 1979) or by utilising their own metabolism (Heinrich 1974) to control body temperature.

Behavioural thermoregulation at low ambient temperatures often involves exothermic regulation by microhabitat selection. An example of this involves the southern corn rootworm, *Diabrotica undecimpunctata*, from North Carolina, which elevates body temperature above ambient air temperature by basking in direct sunlight on clear autumn and winter days (Meinke and Gould 1987). On cloudy days basking behaviour was not exhibited and the body temperature was more highly correlated with air and ground temperature.

Physiological thermoregulation, which makes use of the insect's own metabolism, is at its most well developed in bees. Colonies of *Apis mellifera* (Hymenoptera: Apoidea) form clusters in response to low temperatures occurring in the early part of winter, and maintain the temperature at the edge of these clusters above 9 °C (Seeley and Heinrich 1981). Honey is used by the bees as an energy source in the generation of heat, and it is thus important that the bees have an adequate store available for use during the winter.

Using a cold environment

In a few species a cold winter environment has been used to enhance survival. For example, the ability of the Mediterranean fruit fly, *Ceratitis capitata* (Diptera: Tephritidae), to survive food and water deprivation is enhanced by a cold environment (Carante and Lemaitre 1990).

Snow and overwintering sites above the soil surface

The insulating properties of snow on overwintering sites above the soil surface have been noted by many authors (Fuller *et al.* 1969; Pruitt 1979; Werner 1978), and its importance to overwintering insects was strikingly demonstrated by Shorthouse *et al.* (1980). These authors studied three species of cynipid gall wasps (Hymenoptera: Cynipidae) in which prepupae were overwintered at various heights above the ground. Their results showed that autumn temperatures were similar in leaf litter and at heights of 50 and 150 cm above the soil surface. However, after snow fell those sites covered by snow maintained considerably higher temperatures than those without snow (Fig. 2.5). Most of the prepupae that were continually buried beneath snow survived winter but those at heights subject to fluctuating snow depths suffered a considerably greater mortality.

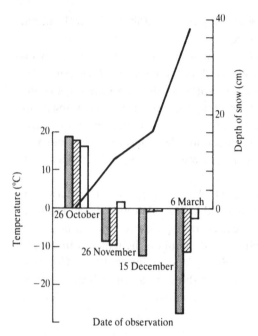

Fig. 2.5. Temperatures and depth of snow on three days during autumn and winter in Ontario (———— snow depth; ▦, ▨ and □ temperature at 150 cm, 50 cm and in the leaf litter, respectively) (after Shorthouse *et al.* 1980).

A similar result was obtained in studies on the overwintering survival of the pine sawfly, *Diprion similis* (Hymenoptera: Diprionidae), in Canada. During the extremely severe winter of 1933–4 the overwintering cocoons of this insect were studied at various heights above the ground. Snowfall was very heavy (2.9 m) and survival varied according to the height above ground. No cocoons above 1 m from the ground survived but 25 per cent of cocoons on the ground were killed (Monro 1935).

Snow depth is also an important environmental factor in the survival of overwintering spear-marked black moth, *Rheumaptera hastata* (Lepidoptera: Geometridae). Pupae of this moth spend the winter in leaf litter where the temperatures are directly related to snow depth at any given air temperature (Werner 1978).

Even when insects overwinter in a protected environment snow cover can still be critical. The larvae of the tortricid moth, *Endothenia daeckeana* (Lepidoptera: Tortricidae), overwinter in the pithy stalks of their host, the Northern pitcher plant, *Sarracenia p. purpurea*. During September, the larvae leave the ovaries where they have spent the summer and tunnel down the flower stalks. They then spend the winter in the now hollow stalk, where they are protected from extreme cold by the snow cover that completely blankets stalks during winters of average or greater than average snowfall (Hilton 1982).

Vegetation

The thermal environment above ground is further modified by vegetation. For example, in forests, shelter from the sun, heat loss by evapotranspiration, blanketing at night, reduction of wind speed and the impeding of vertical air movement all influence the temperature. The overall effect is to lower daily maximum temperatures and raise minimum temperatures. In summer this effect is particularly noticeable in temperate forests at sea level, when mean monthly temperatures may be 2.2 °C less than in open country. In winter the equivalent difference can be as little as 0.1 °C (Barry and Chorley 1976). Thus, although it is difficult to generalise, differences in temperature in temperate forests are most noticeable in the summer months. In winter, forests affect insects more by providing sheltered overwintering sites that offer protection from wind, predation and other factors, than by raising ambient temperatures. Protection from such factors can be extremely important to overwintering insects as is well illustrated by a study of the European corn borer, *Ostrinia nubilalis* (Lepidoptera: Pyralidae), where survival of overwintering larvae was 50 per cent lower when they were not protected from predation by crows from a roost 26 km from the area studied (Quiring and Timmins 1988).

Microclimate differences due to low vegetation cover have been clearly illustrated in orchards in West Virginia (Sharrat and Glen 1988a,b). This study compared microclimate factors on radiation frost nights in spring, between two 1.4-hectare peach plots. One had a complete grass cover and the other alternating 3 m soil and grass strips, with coal dust applied to the soil strip of the tree row. It was found that the net radiative flux on radiation frost nights from the coal dust plot was lower (greater radiative loss) than from the grass plot. However, differences of only 0.5 and 1 °C, respectively, were found in bud and air temperatures between plots, as little of the energy liberated by the soil was intercepted due to the fairly open canopy of the orchard. It was calculated that if all the energy released from the soil was trapped in the canopy then a potential 2–3 °C difference in canopy temperature could be realised between the two plots in the study. However, such differences in canopy temperature would be site specific, as a result of factors such as aspect, slope, relative position on the slope and different types of vegetation cover, as discussed elsewhere in this chapter. In practice it is likely that vegetation cover has little effect on overwintering survival.

Selection of the overwintering site

As the earlier sections of this chapter show, the overwintering site is of great importance in determining the survival of both individuals and populations over the winter months. It is thus in the interest of the insect to have an appropriate suite of responses to available cues, which will enable the best overwintering site to be selected. In this section we examine the cues available and the responses shown by a variety of different insects and relate these to the sites chosen for overwintering.

One of the problems inherent in studying these cues and responses by simple observational means is the difficulty in distinguishing between mechanisms eliciting the same responses, e.g. a digging reflex, which can be brought about either by a negative response to light or by a positive response to gravity. The following sections will highlight these and other problems.

Phototaxis

A number of insect species show seasonal changes in their response to light and the seasonal development of negative phototaxis controls entry into overwintering sites such as hollows, caves or the leaf litter. These responses are sometimes strengthened at low temperature as in the carabid beetle, *Agonum assimile* (Payk) (Coleoptera: Carabidae) (Neudecker 1974). The fly, *Pyrellia serena* Meigen (Diptera: Muscidae), overwinters as an adult in caves in Europe and the United States and the development of a negative phototaxis as winter approaches means that it is more likely to find itself in a dark cave than exposed in the outside environment. As it is usually found in aggregations, it was hypothesised that a thigmotactic response is also present. However, laboratory experiments showed no sign of such a reaction. The negative phototaxis was brought about by a simple lowering of temperature in the laboratory (Holmquist 1928).

In some insects negative phototaxis is accompanied by a digging reflex. The coccinellid *Coleomegilla fusilabris* (Mulsant) (Coleoptera: Coccinellidae) overwinters in the litter layer of woodland. As winter approaches, it becomes negatively phototactic and if placed on a yielding surface will begin to dig (Park 1930). Presumably this is the same reflex observed in the prepupal stage of *Panolis flammea* (D & S) (Lepidoptera: Noctuidae), which, when dropped on to a peat/needle litter layer, promptly burrows downwards (Leather and Brotherton 1987) and, on location of a suitable overwintering site, pupates.

Another aspect of negative phototactic responses is the attraction of some insects prior to overwintering to black objects. The openings of caves and crevices can appear as black surfaces against light backgrounds and thus elicit the negative phototactic response of many insects. The Compton tortoiseshell, *Nymphalis vau-album* D & S (Lepidoptera: Nymphalidae), is reported to respond to openings in this way in the United States (Proctor 1976) and the moths *Scoliopteryx libatrix* (L.) (Lepidoptera: Noctuidae) and *Triphosa dubitata* (L.) (Lepidoptera: Geometridae) also fly towards black surfaces in the autumn (Tercafs and Thines 1973). This response varies with temperature in *T. dubitata*, being more marked at low temperatures (Tercafs and Thines 1973). Flying towards a black surface at the onset of winter causes the moths to aggregate in caves or similar locations. The response disappears with the ending of the winter conditions, although it is not coincident with a response to clear light surfaces.

Some insects show positive phototactic responses as winter approaches. Adults of the convergent ladybird, *Hippodamia convergens* Guerin-Menevill (Coleoptera: Coccinellidae), are attracted to clearings on slopes exposed to afternoon sunlight and concentrate in these spots until the sun sets, thus forming an aggregation (Hagen 1962). A similar response is shown by a related species, *Hippodamia quinquesignata* Kirby (Coleoptera: Coccinellidae), which move from their feeding sites on the prairie to aggregate in their overwintering hibernation sites under rocks and debris on west-facing slopes (Harper and Lilly 1982).

This negative phototaxis and positive thigmotaxis will tend to direct insects into overwintering sites where they are protected from larger predators (e.g. under stones, in small hollows, under bark, etc.) as well as from the vagaries of the winter environment.

Thigmotaxis

Many insects overwinter in large aggregations, either as protection against inclement weather e.g. the monarch butterfly, *Danaus plexippus* L. (Lepidoptera: Danaidae) (Calvert *et al.* 1983) or to facilitate mating in the spring, e.g. coccinellids (Hagen 1962). As seen earlier, aggregations can be caused by phototactic and hygrotactic responses. However, one of the more common causes of aggregation is thigmotaxis – the response to touch or contact. The cereal thrips, *Limothrips cerealium* Hal. (Thysanoptera: Thripidae), a pest of cereals in Britain, overwinters in the crevices of tree bark. Lewis and Navas (1962) showed in the laboratory that this was not a phototactic response but that the thrips were entering deep, narrow crevices in response to tactile stimuli. Furthermore, they were able to show that, as a result of this behaviour, their overwintering sites were confined to a limited range of widths (0.23–0.28 mm) because they were unable to fit into smaller crevices and wider crevices made it difficult for them to achieve maximum contact with their surroundings. The ladybird beetle, *Coleomegilla maculata* (DeGeer) (Coleoptera: Coccinellidae), overwinters in aggregations in the forest floor. It feeds in the forest foliage and, as autumn approaches, moves downwards (probably due to negative phototaxis and, perhaps, a geotactic response (Hodson 1937)), but once in the forest floor, they become positively thigmotactic.

Hygrotaxis

Many insects respond to humidity gradients, e.g. the alfalfa weevil, *Hypera postica* (see p. 23), although it is often in combination with other responses. The ladybirds *C. maculata* and *H. convergens* both show responses to humidity once their overwintering site has been roughly located (Hodson 1937; Hagen 1962). Bumble bee queens, *Bombus agrorum* Fabr., *Bombus hypnorum* L., *Bombus pratorum* L. and *Bombus terrestris* L. (Hymenoptera: Apoidea), choose their overwintering sites on the basis of texture and humidity (Pouvreau 1970). They prefer well drained ground, shaded from direct sunlight and preferably with a north-west exposure (Alford 1969).

Thermotaxis

Temperature is an important factor influencing the choice of overwintering site of many insects. A fascinating example of this is shown in some arctic mosquitoes. *Aedes impiges* (Walker) and *Aedes nigripes* Zett (Diptera: Culicidae) overwinter as eggs. The adults oviposit on the ground surface near pools but only moist sites and sites that are not shaded by vegetation are chosen. In addition the eggs are laid only on slopes that face the sun during the warmest time of the day and, in the main, on those slopes that are more or less normal to the sun's rays. The position of the eggs ensures that the majority will be covered by water in the spring thaw and the choice of the warmest sites means that these areas thaw out first. The eggs thus hatch at the earliest time possible to take advantage of the short arctic summer (Corbet and Danks 1975). The adult mosquitoes respond to the humidity but, to a larger extent, to the temperature of the microclimate to achieve this precise selection of overwintering site.

Multitactic responses

Most insects respond to more than one stimulus to locate a suitable overwintering site. A good example is provided by the larvae of the mayfly, *Cloeon dipterum* L. (Ephemeroptera: Baetidae), which overwinter in anoxic ice-covered ponds in Sweden. In this habitat the most aerated conditions occur immediately below the surface of the ice; this is also the coldest area of the pond. Once the pond freezes, and when the water at deeper levels becomes anoxic, many of the *C. dipterum* larvae migrate to the underside of the ice and remain there for the winter. In a series of elegant experiments, Nagell (1977) showed that, under aerated conditions, larvae moved towards the darker part of the light gradient and, sub-sequently, in anoxic conditions they became positively attracted to the light. A thermotactic response was also detected whereby in aerated conditions higher temperatures were preferred and colder conditions were preferred in anoxic conditions (Fig. 2.6). Thus, by utilising the natural gradients of light (increasing intensity towards the surface) and tempera-ture (decreasing towards the surface) *C. dipterum*, by its sophisticated shift in behaviour at the onset of anoxia, ensures its survival over the winter.

The alfalfa weevil, *Hypera postica* (Gyllenhal) (Coleoptera: Curculio-nida) overwinters in adult aggregations in small holes (Pienkowski 1976). It shows a direct response to humidity (hygrotaxis), having extremely sensitive hygroreceptors, which can detect differences in relative humidity of only 5 per cent. It also shows a negative response to light and a positive thigmotaxis. Its extremely sensitive response to humidity gradients was strikingly demonstrated by the discovery of a massive aggregation of *H. postica* at a reservoir near Fort Collins, Colorado, where the insects had aggregated in a narrow band under small rocks about 1.5 m above the surface of the water across the entire western face of the dam (Simpson and Welborn 1975).

All of these responses result in a common end – the selection of the optimal overwintering site. Many of the responses ensure that the insect

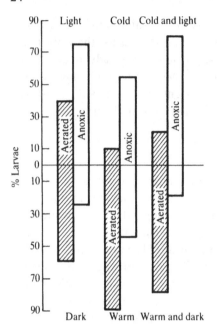

Fig. 2.6. Distribution of larvae of *Cloeon dipterum* along light and temperature gradients after 240 h in anoxic and aerated conditions. ▨ aerated conditon, ☐ anoxic condition (redrawn from Nagell 1977).

is protected not only from the environment but also from predators. Other responses, e.g. aggregations, ensure that the insects, particularly highly vagile species, are able to mate before their spring and summer dispersions. As the ultimate aim is survival of the species over the winter period, it is not surprising that, although the tactics are many, the sites selected are specific and limited.

3

The stimuli controlling diapause and overwintering

Introduction

Insects show a range of responses to the deteriorating environmental conditions that occur as winter approaches. These responses include migration, diapause, polymorphism and dormancy. Whichever response or set of responses the insects (and mites) employ it is important that the developmental stage in which they overwinter (see Chapter 5) is reached at precisely the correct time of year, and they use a variety of seasonal stimuli to keep their life cycles synchronised with the environmental conditions they are experiencing. This chapter considers the abiotic and biotic stimuli that insects use both to detect the forthcoming harsh conditions of winter and to determine the duration of the overwintering state.

Although only a minute proportion of the species that overwinter in temperate areas have been studied, the degree to which individual species rely on seasonal cues indicating the onset of winter already shows a wide variation from complete environmental control at one end of the continuum to genetic control at the other (Tauber *et al.* 1986). The subject is further complicated by the range of token stimuli employed. Photoperiod and thermoperiod frequently act as cues for the control of overwintering but other factors such as moisture, nutrition and, in parasitoids, the physiology of the host, have also been shown to have an influence. The way in which these stimuli act also varies between species. For example, some insects respond to the relative duration of the two phases of thermoperiod (see p. 36), whereas in others the lowest temperature reached in relation to a critical level is important. In addition, in many species the effect of two or more stimuli interact and this possibility must be included in any study of the induction, maintenance or termination of the overwintering state. Differences in the response to overwintering cues between insects of the same species from different geographical areas have also been reported. In many insects only one developmental stage is sensitive to the token stimuli and this stage may not be the same as the resistant overwintering stage. Finally, in some social insects, not all individuals have to be exposed to such cues before they begin to prepare for harsh conditions, as the information appears to be communicated between members of a colony.

Thus the range of adaptations of insects to environmental conditions indicating the onset of winter is complicated and a fuller account of the subject than is appropriate here can be found in the monograph on seasonal adaptations in insects by Tauber *et al.* (1986). However, despite the wide differences that occur between species, some general patterns can be discerned.

The induction of the overwintering state

Although insect diapause is genetically controlled the environmental conditions that the insect experiences often determine if and when diapause occurs and the extent of the diapause. Typically, factors that regulate diapause induction take their effect during a defined period of an insect's development, when it is sensitive to such cues, and diapause occurs either in the same or subsequent developmental stages or, in some species, in a subsequent generation.

Taylor and Spalding (1988) have proposed five distinct pathways that are commonly followed by arthropods in determining when the diapause occurs:

1. The first is where the sensitive period and diapausing stage both occur in the juvenile instars of the same generation.
2. The second is when the adult diapause stage is preceded by a determinate sensitive period, resulting in adults being committed to either reproduction or diapause prior to reproducing.
3. The third case also involves an adult diapause stage but with an indeterminate sensitive stage that extends into the reproductive adult stage, thus allowing adults to reproduce before entering diapause.
4. In the fourth case the parental generation determines diapause induction in the next generation and all offspring diapause or all are non-diapausing.
5. The final case is similar to the fourth but where both diapausing or non-diapausing offspring are produced.

Functions that assign a fitness to the switching time (the point at which the decision to diapause is made) have been developed for each of these pathways (Taylor and Spalding 1988). In each case these illustrate that there is an optimal time at which individuals should enter diapause or produce diapausing offspring, and so the timing of the response to environmental stimuli indicating the onset of harsh conditions appears to be critical in many species.

Abiotic cues

Despite this diversity of possible pathways for diapause induction, the same few regulatory environmental cues are important in all the species that have been studied. Three abiotic (photoperiod, temperature and moisture) and two biotic (nutrition and crowding) cues are known to be important, and of these the first two abiotic cues predominate in most insects. Many species detect the onset of harsh conditions more reliably by responding to more than one of these cues, thus attaining a closer adaptation to the environment.

Photoperiod

Photoperiod is one of the most reliable guides that is available to insects to distinguish forthcoming changes in seasons. Changes in day-length follow a regular seasonal pattern, which is more reliable than the patterns of other features of the environment. The day-length pattern is particularly well defined in areas distant from the equator (see p. 6). Insect species use incoming radiation in several ways to regulate their life cycle in relation to the environmental conditions that govern population development. As well as responding directly to day-length, they also have been shown to respond to changes in day-length occurring over short periods of time. This has, in turn, highlighted the importance of spectral sensitivity, especially in species living in areas with a long twilight (Lees 1955), and also of the responses to photoperiods above and below those that are important in inducing diapause. Two daily phases are distinguished in studies of the effect of photoperiod on insects, a light phase (photophase) and dark phase (scotophase).

The effect of photoperiod on induction of the overwintering state   Insects often respond to either a short photophase or long scotophase as a token stimulus indicating the deterioration of weather conditions as winter approaches. Many aphids rely on parthenogenetic viviparous reproduction to sustain their colonies through most of the summer and

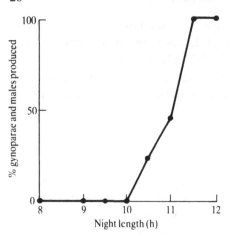

Fig. 3.1. The production of sexual morphs by *Rhopalosiphum padi* when it is reared at different night-lengths at a constant 14 °C (after Dixon and Glen 1971).

only produce sexual forms (males and oviparae) in autumn. These sexual forms mate and lay the eggs that are the overwintering stage of many species. The oviparae are produced by another special morph that only occurs in autumn, the gynopara. The switch to the production of males and gynoparae has frequently been associated with short day-length, although it is actually triggered by long scotophase (Lees 1973). The bird cherry-oat aphid, *Rhopalosiphum padi*, is an example of an aphid in which the production of these two morphs is strongly influenced by the length of the scotophase. When reared in the laboratory under a series of constant night-lengths (Dixon and Glen 1971) it was shown that at 14 °C they were produced when night-length exceeded 10 hours (Fig. 3.1). The production of both oviparae and males of the pea aphid, *Acyrthosiphon pisum*, is also influenced by length of scotophases (Lees 1989). However, this study showed that all ovipara-producing aphids switch spontaneously to virginopara production part-way through the progeny sequence, irrespective of photoperiod, and this may confer the ability to overwinter parthenogenetically in mild winters. Thus the pea aphid shows a variable strategy, which depends on environmental conditions both at the time of production of the sexuals and afterwards.

As in these examples, laboratory experiments investigating the effects of abiotic stimuli on the induction of overwintering states in insects are often conducted by exposing insects to unchanging photophase/scotophase ratios. This allows the convenient analysis of the data by plotting

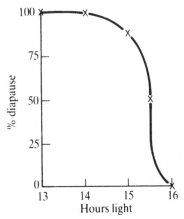

Fig. 3.2. The effect of photoperiod on the induction of diapause in *Cotesia* (*Apanteles*) *rubecula* at 20 °C (after Nealis 1985).

response curves, which usually show a characteristic shape, and by the determination of a critical photoperiod. The parasitic braconid wasp *Cotesia rubecula*, for example, responds to deteriorating environmental conditions by entering diapause. Studies conducted in constant temperatures demonstrated a clear photoperiod response in the induction of diapause (Nealis 1985) and the response curve that can be plotted from these results illustrates the typical shape (Fig. 3.2). The critical photoperiod is defined as that photoperiod that causes 50 per cent of the insects in the experiment to enter diapause, and in this case is approximately 15.5 hours. Critical photoperiods for diapause induction can vary between geographical location (see p. 67) and species. The squash bug, *Anasa tristis*, undergoes reproductive diapause from late summer to spring and, in the laboratory, diapause was induced in 100 per cent of adult females reared under photoperiods shorter than 14 hours (Nechols 1988). The critical photoperiod for diapause fell between 14 and 14.5 hours, a range that compared closely with the prevailing day-length when about 50 per cent of the adult population enters diapause in the field. Reproductive diapause in the collembolans *Orchesella cincta* and *Tomocerus minor* is induced by photoperiod and temperature. Critical photoperiod estimated from field data lies between 11 and 13 hours and induction relies on field temperatures of between 10 and 16 °C (Woude and Verhoef 1988).

Insects that have an identical critical photoperiod can still differ in their response to day-length as the steepness of the response curve may not be similar. Lees (1968) suggested that those species that were under greater pressure from natural selection for timing of the induction of the

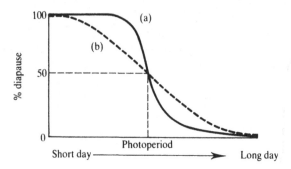

Fig. 3.3. Two hypothetical insects under (a) strong and (b) weaker natural selection for timing of the induction of the overwintering state.

overwintering state show a steeper curve through the critical photoperiod (Fig. 3.3). Many other examples of photoperiodic response curves for the induction of the overwintering state have been reported (e.g. by Saunders 1965; Peterson and Hamner 1968; McNeil and Rabb 1973; Anderson and Kaya 1974; Caldwell and Wright 1978; McLeod *et al.* 1979; Wylie 1980; Hegdekar 1983; Kudo and Kurihara 1988) and a fuller account of some of these studies can be found in Beck (1980) and Tauber *et al.* (1986). As will be seen later, insects often have different critical photoperiods for the induction, maintenance and termination of diapause.

   Tauber *et al.* (1986) point out two limitations of the use of critical photoperiods. First, they do not indicate the intensity of the diapause induced by different photoperiods. For example, the two-spot ladybird, *Adalia bipunctata*, undergoes a shorter diapause if it is induced by day-lengths close to the critical photoperiod than if induction is the result of day-lengths that are well below the critical level (Obryki *et al.* 1983). Secondly, critical photoperiods are based on constant cycles of light and dark and do not take any account of the effect of the changes in day-length to which many insects also respond.

The effect of variable photoperiod   The effect of sequential changes in daylength on induction of overwintering has been studied for several insects. As with other aspects of the influence of photoperiod, different species show different responses to changing day-length. For some, only the precise day-length in relation to the critical period is important and changing day-length has little effect on diapause induction (e.g. the fruit tree red spider mite, *Panonychus ulmi*). In others, the exact day-length is less important than one that is constantly changing. Often the range of the change in day-length has to include a critical photoperiod if it is to induce diapause in a large proportion of individuals in the population, although this is not always the case. The proportion of diapause eggs laid by the grasshopper *Chorthippus bornhalmi* has been associated in laboratory studies with both the length of constant day-lengths (e.g. 80 per cent diapause eggs are produced at light:dark (LD) 14:10) and with increases and decreases in day-length (Ingrisch 1987). There are even a few insects that do not have to encounter day-length of a critical photoperiod to enter diapause. The lacewing *Chrysopa carnea* will respond to a rapid reduction of day-length even if this does not reach the critical day-length (Tauber and Tauber 1970).

The geographical variation noted in several diapause responses (see p. 67) has also been shown in responses to variable photoperiods. A study of the comma butterfly, *Polygonia C-album*, populations (Nylin 1989) originating from Sweden and England has shown that at 20 °C constant day-length was ineffective in averting the production of the diapausing morphs in Swedish strains whereas non-diapausing morphs were obtained if day-length were increased during larval development. English forms showed a critical constant day-length for both morph and diapause induction of between 12 and 18 hours. However, decreasing day-length above the critical day-length early or late in larval development also resulted in production of the diapausing morph, a similar result to that reported for *Chrysopa carnea* (Tauber *et al.* 1970).

Interaction of photoperiod with other cues to induce diapause

Induction of the overwintering state frequently (or usually) results from the combined effects of several cues rather than from a single cue. Indeed, several authors have suggested that an interaction between photoperiod and other environmental cues often provides a safety device in diapause determination (Danilevskii 1965; Lees 1968; Beck 1980; Takeda and Chippendale 1982). The most common factor interacting with photoperiod is temperature, which typically acts through three mechanisms: first, by setting upper and lower thermal limits between which insects respond to photoperiod stimuli, and second, by altering critical photoperiods. In addition, conditions of temperature experienced between the perception of the diapause-inducing stimulus and the start of the diapause can alter or cancel the diapause-inducing effect of photoperiod or can affect the depth or duration of diapause. One example of the first mechanism involves the southwestern corn borer, *Diatraea grandiosella*, where high temperatures can suppress, and temperature cycles can influence, photoperiodic determination of diapause (Chippendale and Reddy 1973; Chippendale *et al.* 1976; Takeda and Chippendale 1982). The alteration of critical photoperiods is displayed by the codling moth (*Cydia pomonella*) where the critical photoperiod for diapause induction is temperature-dependent (Garcia-Salazar *et al.* 1988). Seasonal temperatures preceding the initiation of diapause can shift the critical photoperiod towards a shorter (by higher temperatures) or longer (by lower temperatures) day-length. Thus at 27 °C the critical photoperiod for 50 per cent induction of diapause was LD 13.8:10.2 hours, while at 17 °C it was LD 15:9 hours. Higher temperatures will therefore either diminish or avert the photoperiodic response of the larvae. Investigating the seasonal phenology of the codling moth, Riedle and Croft (1978) found that the critical photoperiod was not constant for a population but varied between years, probably due to the modifying effect of prediapause temperatures.

More generally, investigations involving several long-day species have shown that high temperatures and long days often act together to avert diapause whereas low temperature and short days act to induce it (Danilevskii 1965). Temperature can also affect the induction of diapause by shortening (high temperature) or extending (lower temperature) the duration of the photosensitive period (Danilevskii *et al.* 1970).

Several examples of the interaction of photoperiod with a cue other than temperature to induce the overwintering state have also been described. In *Chorthippus bornhalmi* the induction of egg diapause is primarily related to photoperiod (see p. 39). However, under certain conditions the percentage of diapausing eggs was found to be higher in older females, indicating that maternal age may also affect diapause induction (Ingrisch 1987). Other examples are highlighted in later sections.

Temperature

Temperature is the second abiotic stimulus that has been found to be of major importance in the induction of insect overwintering. Environmental temperatures go through daily cycles in which daytime temperatures are higher than night-time temperatures. Daily thermoperiods, like photoperiods, vary seasonally and, accordingly, can be used to indicate oncoming harsh conditions. However, as thermoperiod is subject to a greater degree of fluctuation between years than photoperiod, it is less reliable as a seasonal indicator and is infrequently found to be the sole factor controlling the appearance of the overwintering state. When it is the major regulator the insects concerned often live in a specialized environment, such as a subterranean habitat. It is more often found to act by modifying or reinforcing the effect of other stimuli, such as photoperiod, as was seen on page 31.

Daily thermal cycles can be divided into two phases, night-time temperatures, or cryophase, and daytime temperatures, or thermophase. In many regions thermophase and cryophase are coincidental with photophase and scotophase, but Beck (1983a) suggests that thermoperiod and photoperiod are not simple alternatives, as there are noticeable differences between photoperiod and thermoperiodic responses. The interrelationship between the two can in many cases be important in diapause determination.

Temperature as a token stimulus indicating the onset of winter   The study of insect thermoperiodism in relation to overwintering is complicated by two factors. First, insects are poikilothermic, and temperature influences their rate of development, the number of periodic cycles to which the insect is exposed during its various stages of development and, sometimes, the nature of developmental determination (Beck 1982).

Secondly, apart from a few rare cases where photoperiod has been shown to play no role in the induction of the overwintering state, photoperiodic and thermoperiodic responses can be difficult to separate. The overriding effect of photoperiod can be controlled in laboratory experiments by keeping insects under conditions of constant darkness, while subjecting them to varying temperature regimes (Danilevskii 1961; Beck 1962; Saunders 1973; Chippendale *et al.* 1976). Few studies have been made using this technique but the information that is available indicates that the effect of thermoperiod can only be demonstrated under conditions of constant darkness. The fullest investigation that has been published is Beck's study of the effect of temperature on the induction of diapause in the European corn borer, *Ostrinia nubilalis* (see Beck 1983a for review).

In one unusual case an alternative approach has been used to investigate diapause induction. The dipteran *Chymomyza costata* enters diapause at the third larval instar in response to photoperiod and temperature (Enomoto 1981; Riihimaa 1984) but low temperatures have been shown to induce diapause independently of photoperiod (Riihimaa and Kimura 1988). A strain of this species that was established by selection (for a single generation only) lacked a photoperiod response but still responded to low temperatures of 11 °C by entering diapause.

A problem encountered in some studies has been the differing reactions that can be obtained when the same species is subjected to constant or variable temperatures. Such differences were investigated by Hegdekar (1977, 1983) using the bertha armyworm, *Mamestra configurata*, which has a photoperiodic response curve of the long-day type that is modified by temperature. At constant temperatures of 23 °C a cycle of 12L:12D resulted in a 100 per cent diapause response, whereas the proportion of pupae entering diapause dropped at higher temperatures and at 27 °C no diapause was recorded. At 17 °C, 100 per cent diapause occurred irrespective of photoperiod. However, if the insects were reared under daily fluctuating temperature regimes of 25 and 10 °C (mean = 17.5 °C) the estimated critical photoperiod was 13.5 hours and under similar fluctuating conditions in nature the critical photoperiod was in excess of 16 hours.

Most studies have involved subjecting insects to constant temperatures (for examples see p. 33), but some investigating the effect of thermoperiod have been published. Although thermoperiods often do not induce diapause, they strongly affect the timing of induction. Natural temperatures vary in a cyclic fashion during the day, usually rising from dawn to a maximum early in the afternoon (thermophase) then declining to a minimum (cryophase), which often occurs just before dawn. Three sources of variation in thermoperiod (Beck 1983a) may be used to detect the onset of winter conditions. These are:

the temperatures of the two phases
the duration of the two phases
the magnitude of the difference between the temperature of the two phases.

As relatively little work has been done examining the effect of temperature on its own on the induction of diapause, only a few examples of its effect in the absence of other stimuli are available to illustrate the use of these sources of variation.

The temperature of the two phases   The importance of the environmental temperatures experienced in the induction of diapause has been demonstrated in several insects. Using a constant thermoperiod of 12 hours cryophase/12 hours thermophase in constant darkness, Beck (1982) varied cryophase temperatures while keeping thermophase temperature constant. With a thermophase temperature of 25 °C he showed that cryophase temperatures below 15 °C induced diapause in the European corn borer in almost all the insects tested (Fig. 3.4). Cryophase temperatures of 20 °C resulted in virtually no diapause. Other similar experiments were carried

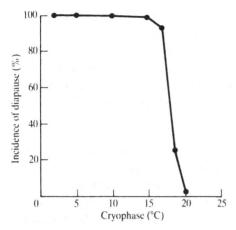

Fig. 3.4. Induction of larval diapause in the European corn borer, *Ostrinia nubilalis*. Thermoperiod = 12C:12T, thermophase temperature = 25 °C, cryophase temperature varies between 2 and 20 °C (after Beck 1982. Reprinted from *Journal of Insect Physiology* with permission from Pergamon Press).

out using thermophase temperatures of 30 and 35 °C. The results indicated that any cryophase exceeding 18.5 °C was ineffective in inducing diapause irrespective of the thermophase temperature, and cryophases lower than 17 °C resulted in high incidences of diapause. He concluded that there was a response threshold below which cryophase temperatures must fall in order to induce the diapause response, which he estimated as being close to 17.5 °C.

Response thresholds for other species have been estimated from published data and include 19 °C for *Pieris brassicae*, 18 °C for *Diatraea grandiosella*, 16 °C for *Praon exsoletum* and 13 °C for *Nasonia vitripennis* (Beck 1983a). In all cases the estimated thermoperiodic response threshold is higher than the expected developmental threshold.

In the few cases where photoperiod has been shown to play little or no role in preparing insects for the onset of winter conditions the role of temperature has been more readily defined. In these situations thermoperiodic induction of diapause can be completely dependent on the cryophase temperature being below a critical level. For example, the dipteran *Chartophila brassicae* undergoes pupal diapause in the soil, where temperature is less variable than in the air above (see p. 10). Diapause is induced by low soil temperatures of about 15 °C (Missonier 1963).

Instead of overwintering in the egg stage as described previously (see p. 27) some aphids remain as viviparous females throughout the winter. Amongst these is the lettuce root aphid, *Pemphigus bursarius*, which responds to environmental cues indicating the forthcoming harsh conditions, by forming an overwintering morph called a hiemalis. The hiemalis has poorly developed gonads and a well-developed fat body (Judge 1967; Sutherland 1968) and can survive without a host plant for up to 48 weeks at a temperature of about 3 °C (Dunn 1959). The hiemalis of *P. bursarius* develops in response to low temperatures, 95 per cent of individuals developing into this morph at 10 °C.

The duration of the two phases   The exact temperature reached during the cryophase is not the only factor which is important in the induction of the overwintering states. The duration of the period that the cryophase remains below the critical temperature can also be important in some insects. Again the European corn borer provides an example. A similar level of larval diapause is induced by a photoperiod of 12 hours day and 12 hours night, or if the insects are subjected in constant darkness, to a 12 hour cryophase at 10 °C with a thermophase of 12 hours at 31 °C (Beck 1962).

Thermoperiodic response curves similar to those for photoperiod have been described for some species, including the corn borer. In this species the incidence of diapause is proportional to the length of the cryophase (Beck 1983b). Using a cryophase of 15 °C and a thermophase of 30 °C, the thermoperiodic response curve showed a critical cryophase (causing 50 per cent of insects to enter diapause) of about 9.5 hours (Fig. 3.5a) and all the larvae entered diapause if it exceeded 12 hours (Beck 1982).

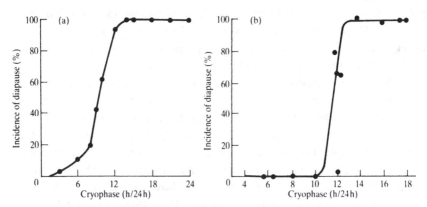

Fig. 3.5. Thermoperiodic response curves for (a) The European corn borer (Beck 1982. Reprinted from *Journal of Insect Physiology* with permission from Pergamon Press) and (b) a parasitic wasp, *Nasonia vitripennis* (redrawn from Saunders 1973; Copyright 1973 by the AAAS).

The proportion of female parasitic wasps of the species *Nasonia vitripennis* producing diapausing broods between the 15th and 17th days of adult life has also been related to the length of daily thermoperiod (Saunders 1973). The wasps were kept in continuous darkness and subjected to constant thermophase and cryophase temperatures of 23 and 13 °C, respectively. The results indicated that long cryophases of between 14 and 18 hours induced almost 100 per cent diapausing broods (Fig. 3.5b) while those of 10 hours or less did not result in any diapause. The sharp discontinuity between these two levels represents the range of effective cryophase lengths including the critical cryophase.

The initiation and pattern of embryonic development in the speckled bush-cricket, *Leptophyes punctatissima*, have been shown to be dependent on temperature (Deura and Hartley 1982). Immediate high temperature facilitated rapid and direct embryogenesis to the diapause stage whereas slightly lower temperatures resulted in a delay before embryogenesis started which, if temperatures were decreased, was greatly extended as an initial diapause. All these responses increased both as a function of exposure time and according to prevailing temperatures at oviposition and, as a result, *Leptophyes* could be an annual or biennial species.

Non-diapause overwintering states can also be affected by thermo-period. Ambient thermoperiod has been found to play an important role in the larval growth and seasonal biology of the black cutworm (*Agrotis ipsilon*), without inducing a state of diapause (Beck 1988). The early instars of this species are unable to survive prolonged exposure to very cool thermoperiods, but third to sixth instars can survive, and simulation of autumnal thermoperiods indicated that mature sixth-instar larvae were unable to pupate until exposed to a more favourable regime. The study showed that larvae of the black cutworm could survive many months of mildly cold conditions in a state on non-diapause dormancy.

The magnitude of the temperature difference between thermophase and cryophase The importance of the magnitude of the difference between the temperatures of thermophase and cryophase has only been investigated in one species. Studies of the European corn borer revealed that, under laboratory conditions, small or large differences between the two tempera-tures appeared to have no effect on the induction of diapause and that the important factor was the duration of cryophase and whether the cryo-phase temperature dropped below the critical response threshold (Beck 1982, 1983a).

Temperature and the production of cryoprotectants   Seasonal variation in cryoprotectant levels is common to many insects (Sømme 1964; Baust and Miller 1970) and the importance of environmental temperature in the control of these seasonal changes has frequently been noted (Baust and Miller 1970; Mansingh and Smallman 1972). In some insects levels of such cryoprotectants as glycerol increase immediately after diapause starts, irrespective of the temperature of the environment (e.g. Chino 1957; Wyatt and Meyer 1959). For many, however, accumulation begins when inactive insects experience colder temperatures (Mansingh and Smallman 1972). A detailed account of the mechanisms of thermal induction of cryo-protectants is reserved for Chapter 4 and no further discussion will be undertaken here.

Moisture

Although photoperiod and temperature may be viewed as the primary regulatory factors affecting the induction of the overwintering stage in insects, another feature of the abiotic environment has also been shown to exert an influence in some cases. Moisture, like photoperiod and temperature, can vary seasonally in temperate regions, although such variations are usually far less reliable as indicators of the onset of adverse conditions. Accordingly moisture appears to be used infrequently as a cue for induction of overwintering in insects, despite the fact that many insects time their activities to take advantage of the seasonal availability of this resource. Few studies have considered the effect of moisture, and the relationships between moisture and diapause regulation that have been described have largely concerned summer rather than winter diapause.

The subject of the effect of moisture on diapause induction or termina-tion is complicated by the requirement of many insects to absorb water in normal development. For example, eggs of the striped ground cricket, *Allonemobius fasciatus*, absorb water during the early stages of embryo-genesis when kept at higher temperatures, but the absorption of water is delayed until later stages at lower temperatures. This difference has been related to a temperature-induced change in the stages of embryonic development at which the eggs enter diapause (Tanaka 1986). Water uptake appears to occur at the stage at which diapause is initiated or maintained but has not been shown to be a factor inducing diapause in this species. Diapause is also known to influence the absorption of water by the eggs of *Allonemobius socius* (Tanaka 1986).

It is often difficult, therefore, to separate and quantify the effect of the intake of water as a stimulus regulating diapause and although the influence of moisture on diapause termination has been investigated more fully (see p. 55), few relationships between moisture and diapause regulation have been clearly demonstrated. However, one or two studies have yielded indications of a moisture effect. For example, Bakke (1963) speculated that moisture was among the four environmental factors acting during the larval development of cone and seed insects on Norway spruce, and affected the induction of prolonged diapause.

Biotic cues

Although the major factors of the physical environment appear in many cases to have an overriding effect on the induction of the overwintering state, they are not the only stimuli that can be used to recognise the onset of harsh conditions. Some factors of the biotic environment also show a degree of seasonal variation that makes them useful as cues that either trigger the production of the overwintering state, or modify the activity of other stimuli. These biotic factors include nutrition, host physiology in parasites (see p. 62) and, perhaps, crowding.

Crowding

As we have seen, photoperiod is a reliable cue indicating oncoming seasonal changes in temperate climates, but at tropical and subtropical latitudes changes in daylength are less pronounced (see p. 6). Other environmental factors gain an increased importance under such conditions and one of the alternative cues that has received some attention is crowding. This factor has also been shown to be involved in the control of summer diapause in temperate insects such as aphids (Dixon 1975) and diapause in stored product insects such as the Indian-meal moth, *Plodia interpunctella* (Bell 1976). The state of diapause that enables the fig moth, *Ephestia cautella*, to survive the period between December and March when the warehouses it infests are emptied of its food source (citrus pulp) is affected by the degree of crowding experience by the insects. An increased incidence of diapause has been linked to dietary residues which are the result of increasing larval densities (Hagstrum and Silhacek 1980). Thus examples of the effect of insect density on diapause induction are usually linked to stored product insects, or diapause that occurs in warm conditions, and it is only rarely considered as an important factor in the production of the true overwintering state, although a few examples are available.

In the sawfly, *Diprion pini*, diapause induction is essentially dependent on photoperiod and temperature, but is also influenced by other factors (Geri and Goussard 1988, 1989a,b; Geri *et al.* 1988). Among these other factors is the number of larvae in colonies (Geri and Goussard 1989b), with larger groups resulting in more diapausing eonymphs. The larvae of the dipteran *Chymomyza costata* also respond to photoperiod, temperature and crowded conditions that retard larval development, by entering diapause (Bottella and Ménsua 1987).

Nutrition – host plant

The food available to insects from their host plants rarely remains constant throughout the year in either quality or quantity. Most annual insect activity cycles are adapted to the availability of food and some insects make use of changes in nutrition to either trigger the induction of an overwintering state or, more often, to modify the effect of other cues (Saunders 1980). Most of the examples illustrating the effect of nutritional regulation of diapause concern summer diapause (Tauber *et al.* 1986), but several accounts of its effect on the induction of hibernal states have been published.

As well as affecting many insects, the short autumn days of temperate regions induce a cessation of growth and dormancy in plants. It has been shown that some insects detect these changes and use them as cues indicating the onset of unfavourable conditions. The aphid *Dysaphis devecta*, for example, relies on the cessation of shoot growth on its host plant to trigger the production of sexual forms (Forrest 1970). It has also been pointed out that root-feeding aphids, which live in continuous darkness and at more constant temperatures than occur in the air above ground level, produce sexual morphs as the plants become dormant (Dixon 1985). In these cases the exact nature of the cue that induces sexual morph production is not known, but in other insects both the quantity and quality of the available nutrition have been used.

The pitcher plant mosquito, *Wyeomia smithii*, spends the egg, larval and pupal stages in the water-filled leaves of the insectivorous pitcher plant, where larvae feed by filtering the particles of organic material which arise from the breakdown of the plants' prey. Winter is spent in diapause, which is both triggered and terminated by photoperiod (Smith and Brust 1971; Bradshaw and Lounibos 1972). This photoperiodic response has been shown to be modified by the amount of food available for the larva (Istock *et al.* 1975). These authors established populations of mosquitoes with a similar initial larval density and each was subsequently fed every 4 days using a standard nutrient suspension. To investigate the effect of

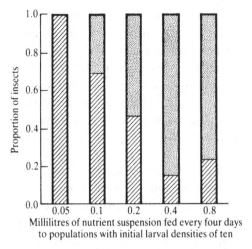

Fig. 3.6. The proportion of pitcher plant mosquitoes that entered diapause when larvae were reared under similar conditions of crowding, photoperiod and temperature but were given different amounts of food. ▨ diapause, ▦ non-diapause (after Istock *et al.* 1975).

food on population development, each population was given one of five different amounts of nutrient solution at each feed. Thus different populations were fed at levels ranging from near starvation to excess food. The proportion of mosquitoes that entered diapause was found to be considerably higher when less food was available than when they were given larger amounts of nutrient suspension (Fig. 3.6).

Food quality is important in the induction of diapause in the Colorado beetle, *Leptinotarsa decemlineata*. Diapause in this pest species is primarily triggered by photoperiod, but the effect of this factor is augmented by the beetle feeding on the older, yellowing leaves of potato plants (Hare 1983). Similarly white-fly of the species *Aleurochiton complanatus* enter diapause more readily if they are feeding on yellowing foliage than otherwise (Müller 1962). Laboratory studies in which the predatory mite *Amblyseius pontentillae* was kept in constant darkness have shown that diapause in this species is thermoperodically induced but that the response is dependent on the presence of vitamin A in the mite's diet (Van Houten *et al.* 1987).

Several factors have been implicated as cues inducing prolonged diapause in cone and seed insects, and among these are moisture (see p. 39) and nutrition (Bakke 1963). The Douglas fir moth, *Barbara colfasciana*, overwinters as a pupa in fir cones. A proportion of these moths emerge as adults in the first spring after pupation occurs, but others remain in diapause and only emerge one or more years later. The proportion that remains in diapause varies from year to year. Incidence of prolonged diapause was related directly to only one weather factor, temperature, but it was suggested that the effect of weather parameters on the host may also affect the induction of diapause through nutritional factors (Hedlin *et al.* 1982). It is interesting to note that studies on the spruce seedworm, *Cydia strobilella*, have shown that insects that entered prolonged diapause were significantly heavier than those that were in diapause for only one winter (Bakke 1971).

Photoperiod, temperature and host plant have been shown to influence the production of diapause eggs in the mite *Petrobia harti* (Koveos and Tzanakakis 1989). When mites were kept at identical diapause-inducing photoperiods and temperatures on detached leaves of plants, between 40 and 90 per cent of females laid diapause eggs on *Oxalis articulata* but only a very low percentage of such eggs were produced on *O. corniculata*. In addition, higher proportions of females laid diapause eggs on leaves from flowering than non-flowering *O. articulata*. However, unlike the Douglas fir moth, laboratory experiments indicated that photoperiod influences diapause induction or aversion by a direct effect on the mites and not by an effect through host leaves.

The nutrition available to the insect not only effects diapause. In some seasonally polymorphic species, such as the psyllid *Psylla pyri*, the production of the winter morph has been shown to be the result of photoperiod, temperature and the vegetative activity of the host trees (Rieux and D'Arcier 1990).

Several other possible influences of nutrition on induction of over-wintering have been reported in the literature, such as an effect of larval diet on the sensitivity to diapause-inducing stimuli in the carrot fly, *Psila rosae*, (Burn and Coaker 1981). However, most authors indicate that further work needs to be done to define more precisely the full extent of nutritional effects on induction of the overwintering state in insects, and it is probable that it is confined to only a small proportion of temperate species.

Maintenance and termination of the overwintering state

The length of diapause is critical for ensuring survival of insects during winter. If emergence from diapause occurs too early then insects will be exposed to the harsh conditions that diapause should protect them from, but if it is too late then they may become out of synchrony with the growth and development of their host plants or animals, or other features of their environment. Insects also have to overcome large year-to-year variations in environmental conditions such as extreme variability in their food supply, and periods of extended dormancy have been linked with risk spreading under such circumstances (see p. 59) (Danks 1987).

Although such periods of extended dormancy have been demonstrated in several studies of, for example, coupled host–parasitoid interactions or species using the same temporally varying resources, many of the theoretical predictions concerning them have not been fully tested (Hanski 1988). However, the length of any diapause is determined by how early it is entered and how late it is completed and, in both cases, a similar range of environmental cues has been found to influence it.

Abiotic cues

Photoperiod

The diverse and dynamic photoperiodic responses related to the induction of overwintering in insects that were discussed on pages 27–32 are also evident in the maintenance and termination of the overwintering state. Many insects appear to have evolved to take advantage of the seasonal progression of photoperiods during winter, and diapause maintenance and termination often involves responses to a series of photoperiods that exert their influence as diapause proceeds instead of to a single critical photoperiod (Tauber *et al.* 1986). Also, the mechanisms and photoperiodic conditions involved are often different from those associated with diapause induction.

A recent study, which illustrates the influence of photoperiod on maintenance and termination of diapause, involved the squash bug, *Anasa tristis*, in Kansas. The insect enters a reproductive diapause by early September (Nechols 1987) and this lasts from late summer to spring. Live adults, collected on various sampling dates between October and March, were subjected in the laboratory to various photoperiods and the length of time before oviposition was recorded. The results (Fig. 3.7) indicated that reproductive diapause in the squash bug was maintained by short day-length and terminated by long day-length. Under natural day-length the duration of diapause became progressively shorter with advancing sampling date. In the field this response results in a prolonged diapause, which may prevent premature postdiapause development in the very variable temperature conditions of Kansas in late winter and early spring

Controlling stimuli

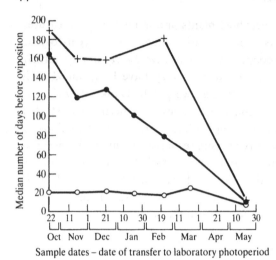

Fig. 3.7. Median number of days for the squash bug, *A. tristis*, to oviposit after transfer on various dates to various photoperiods. ●—● natural day-length, ×—× 11-hour day-length, ○—○ 16-hour day-length (after Nechols 1988).

(Nechols 1988). However, the precise mechanisms and responses underlying diapause in these squash bugs could not be elucidated from the data collected in this study.

More generally Tauber *et al.* (1986) suggest two categories of response to photoperiod in relation to diapause termination and, in so doing, highlight the important differences in the function of photoperiodic maintenance and termination of diapause. In some insects responsiveness to photoperiod is reduced during diapause. In such cases photoperiod may have no direct influence on diapause termination and diapause may either simply run its course to completion or another factor may control its termination. The second category includes those insects where photoperiodic sensitivity persists throughout diapause (as in the squash bug) or where it exerts an influence in the later stages as a diapause-maintaining factor. Here diapause is maintained by photoperiod and a specific photoperiodic cue may be needed to terminate it.

The precise nature of the photoperiodic cues that either maintain or terminate diapause are variable and in many cases have been linked to the local environmental conditions in which the insects live. Thus, in Kansas, the prolonged diapause of the squash bug, which is caused by the photoperiodic response described above, may prevent the insects emerging from diapause in the hazardous early spring conditions that prevail in the area (Nechols 1988). Several other examples of environmental adaptations in diapause maintenance and termination have been reported in the literature.

Reduction in responsiveness to photoperiod as diapause progresses Termination of diapause in some species of insects is not affected by photoperiod and the insects of Tauber *et al.*'s (1986) first category are of this type. The category may conveniently be divided into two types of response. The first is illustrated by the lacewing, *Chrysopa carnea*, where shortening day-length reduces the rate of diapause development but long days do not affect the rate in spring. Thus the duration of diapause is inversely related to day-length but, despite this relationship, diapause is not maintained by day-length, and no particular photoperiod can be associated with its termination (Tauber and Tauber 1973a,b). In the second, which occurs in such species as the dipteran *Tipula subnudicornis*, short days maintain diapause (Butterfield 1976) but again long days have no effect on diapause termination.

Other examples of insects in Tauber *et al.*'s first category include another neuropteran, *Chrysopa harrisii*, the dipteran *Sarcophaga bullata*, the hemipteran *Pyrrhocoris apterus* and the hymenopteran *Tetrastichus julis* (Denlinger 1972; Hodek 1971a; Tauber and Tauber 1974; Nechols *et al.* 1980).

No reduction in responsiveness to photoperid as diapause progresses The second category of Tauber *et al.* (1986) includes insects in which the effect of photoperiod extends throughout diapause and may influence diapause termination as well.

There are two mechanisms by which photoperiod can effect the termination of diapause. The first results from a direct relationship between the rate of diapause development and photoperiod such that diapause develops faster as daylengths increase at the end of winter. Insects such as *Tipula pagana* (Butterfield 1976) and *Chrysopa downesi* (Tauber and Tauber 1976a) are examples. Alternatively, some insects respond to a simple critical photoperiod. Diapause ends when natural daylengths exceed this level.

Diapause in the pitcher plant mosquito, *Wyeomia smithii*, is triggered by photoperiod, although the response may be modified by the nutrition available to the insect (see pp. 40 and 60). Sensitivity to photoperiod persists throughout the winter and diapause is terminated when daylength exceeds a critical photoperiod, which has been measured in Ontario

as between 14.5 and 15 hours (Smith and Brust 1971). However, responses
to critical photoperiods are not always as simple as they may first appear.
These authors also found that overwintering larvae that experienced five
or fewer consecutive long-day cycles remained in diapause when returned
to short-day conditions, but diapause was irreversibly broken in larvae
exposed to eight or more long-day cycles in succession. Responses to
varying sequences of short and long days indicated that overwintering
larvae are more responsive to the diapause-breaking effects of a long-day
photoperiod than to the diapause maintaining effect of short days. When
short and long photoperiods were alternated a large proportion of larvae
(60–77 per cent) broke diapause and pupated (Fig. 3.8) and all the larvae
exposed to repeated sequences of one short day followed by two or more
long days broke diapause (Fig. 3.8a). In contrast, when one long day was
inserted into a sequence of short days, highly variable numbers of larvae
terminated their diapause (Fig. 3.8b). However, the introduction of a
single long day into the sequence caused at least some larvae to complete
their development. These results appear to support the observation of
Beck (1980) that in insects short day effects are more readily reversed
than long day effects.

Other examples of insects in Tauber *et al.*'s (1986) second category
include the orthopteran *Pteromemobius nitidis* and the coccinellid *Adalia
bipunctata* (Obrycki *et al.* 1983; Tanaka 1983).

The effect of changes in day-length   Some insects that do not respond to
critical photoperiods can still use changes in day-length as the cue for
terminating diapause. The category 2 insect *Chrysopa downesi*, which

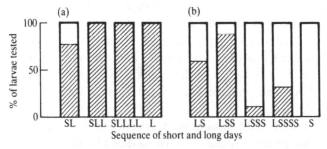

Fig. 3.8. The response of diapausing larvae to varying sequences of short-day and
long-day photoperiods at a constant 20 °C. The sequence of short (S; 12L:12D)
and long (L; 16L:8D) were repeated for 40 days or until 100 per cent pupation.
▨ pupating larvae, □ non-pupating larvae (after Smith and Brust 1971).

must be exposed to short days followed by long days before diapause is terminated, is an example.

Other species require changes in day-length in addition to meeting criteria relating to a critical photoperiod. Larvae of the antlion, *Hageno-myia migans*, resume growth after a period of winter diapause but will not pupate until they have been exposed to a sequence of short days followed by long days. It is essential that day-length in the short days does not exceed 14 hours (Furunishi and Masaki 1981).

Temperature

Temperature is one of the major environmental stimuli controlling the maintenance and termination of diapause. It has been shown to act both in isolation and in conjunction with other cues in the control of intensity, development and termination of the overwintering state. As well as the actual temperatures experienced, the duration of cold exposure has been shown to affect the responses of many species of insect. In others the responses to ambient temperatures are known to vary according to the stage of diapause development. Temperature as a stimulus controlling diapause maintenance and termination is therefore a particularly complex aspect of overwintering biology.

Temperature, diapause maintenance and diapause intensification Several studies have indicated that temperature plays an important role in both diapause maintenance and intensification. Indeed, one of the major functions of temperature in diapausing insects is to maintain the condition, by acting as a regulatory factor on the rate of diapause development.

In many species the temperature range over which diapause development occurs is different from that for non-diapause development. A good example is the western cherry fruit fly, *Rhagoletis indifferens*, in which postdiapause development was not found to occur below 8 °C, whereas the optimum temperature for diapause development was 3 °C (Van Kirk and AliNiazee 1981, 1982). In such insects the low optimum temperatures for diapause development ensure that warm autumn conditions do not result in the resumption of development (Tauber *et al.* 1986). By early or midwinter, when ambient temperatures are usually below those necessary for growth and development, thermal reactions return to non-diapause levels. Subsequent low temperatures can prevent the loss of diapause symptoms, reduce the utilisation of metabolites and synchronise the initiation of postdiapause development with improved environmental conditions. The timing of the return to non-diapause thermal reactions varies between insects – an abrupt change at a particular diapause stage in some and a gradual transition in others.

The difference between the temperature ranges over which diapause and non-diapause development occurs is not always as large as for *R. indifferens*. At the other end of the continuum, as in the orthopteran *Melanoplus bivittatus* (Church and Salt 1952), the two temperature ranges can overlap, although in this particular example the optimum temperature for diapause development was still lower than for non-diapause development. The extreme point of this continuum occurs in those insects where there is little or no difference between the temperatures that allow non-diapause and diapause development to occur. Ten different examples are listed by Tauber *et al.* (1986) and these include *Teleogryllus* spp., *Chrysopa carnea*, *Aedes atropalpus* and *Agromyza frontella* (Tauber and Tauber 1973b; Kalpage and Brust 1974; Masaki *et al.* 1979; Tauber *et al.* 1982).

Where temperature ranges for diapause and non-diapause development are similar, a non-temperature factor or factors may control diapause maintenance. Alternatively in insects such as the dipteran leaf miner, *Agromyza frontella*, (Tauber *et al.* 1982), high temperatures stimulate development and low temperatures can maintain diapause.

Although diapause intensification is frequently controlled by photoperiod, it is also influenced by prevailing temperatures. In several species a longer diapause is entered if a species specific range of temperatures is experienced during a critical period, often in the early stages of diapause.

The precise mechanism by which temperature influences diapause may often be an adaptation to the environment that the insect lives in. More detailed studies of the relationship between diapause and the ecology of particular species of insects will be needed to confirm this.

Temperature as a cue for diapause termination   As was highlighted on page 33, the study of the effects of temperature on insect overwintering is complicated by the difficulty of separating photoperiod effects from temperature effects. The overriding effect of photoperiod on diapause termination can be controlled in laboratory experiments by keeping insects under conditions of constant darkness during diapause while subjecting them to various temperature regimes, and some investigations have made use of this technique. For example studies of the face fly, *Musca autumnalis* (Caldwell and Wright 1978) have shown that this insect requires 4 months exposure to 5 °C under total darkness for the termination of diapause.

Diapause termination can be affected by two features of the thermal environment, the length of low temperature exposure and the actual temperatures experienced. These two factors are not independent of each other as the length of cold storage required to terminate diapause is frequently related to the actual temperature experienced.

A detailed investigation of diapause termination in the encyrtid *Holcothorax testaceipes* has recently been carried out in Southern Ontario (Wang and Laing 1989). *H. testaceipes* is an introduced parasitoid of the spotted tentiform leafminer, *Phyllonorycter blancardella*, an important pest of apples in Ontario. The parasitoid attacks the eggs of the leafminer and subsequently the parasitised larvae will develop through five instars. By the time the host larvae mature, the larval parasitoids have usually consumed all of the host except for the cuticle, which then forms a mummy. For the purposes of this study batches of diapausing insects were held at low temperatures in constant darkness for various lengths of time and were then transferred to higher temperatures and various photoperiods. The end of diapause was defined as the point at which 50 per cent of each batch of diapausing cocoons had emerged. When cocoons were transferred from either 5 or −5 °C to 21 °C at 16L:8D, the length of time before diapause was terminated was inversely correlated with the duration of cold storage. After a period of about 10 weeks exposure, the higher cold storage temperature resulted in considerably shorter delays before 50 per cent emergence from diapause (Fig. 3.9a). Those pupae held at 0 °C also required a shorter exposure to low temperatures to terminate diapause, than those held at −5 °C. Duration of cold storage affected the duration of emergence in *H. testaceipes* (Fig. 3.9b) with longer periods of exposure to low temperatures shortening the period over which emergence takes place. Overall, however, the percentage emergence ranged from 51 to 94 per cent and did not vary significantly with respect to either duration or temperature of cold storage.

Photoperiod also plays some role in the termination of diapause in this species. Although there were fewer days to 50 per cent emergence after longer periods of cold storage regardless of photoperiod, Wang and Laing (1989) found that after various periods of exposure to 0 °C, a shorter period of time passed before 50 per cent emergence in those batches of pupae which were subsequently exposed to photoperiods of 16L:8D at 21 °C than in those exposed to 12L:12D at the same temperature (Fig. 3.9c).

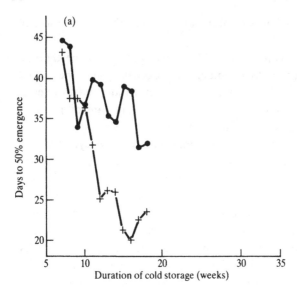

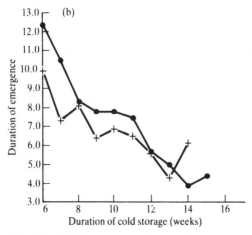

Fig. 3.9. (a) Diapause terminates in *H. testaceipes* after cold storage for varying lengths of time. ●—● −5 °C, ×—× 5 °C. (b) Duration of emergence from diapause in batches of *H. testaceipes* at two different light regimes and after varying lengths of cold storage. ●—● 12L:12D, ×—× 16L:8D (after Wang and Laing 1989).

Both the length of the cold period during diapause and the actual temperature experienced have been shown to effect morphogenesis in several other insects. Pupae of the carrot fly, *Psila rosae*, will terminate diapause after 20 weeks at 1 °C but both spread of emergence of pupae

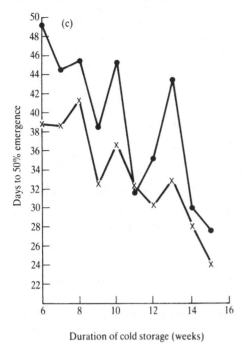

Fig. 3.9. (c) Diapause termination in *H. testaceipes* after cold storage at 0 °C for varying lengths of time and subsequent exposure to two different photoperiod regimes. ●—● 12L:12D, ×—× 16L:8D (after Wang and Laing 1989).

and mean time to emergence decrease with increasing duration of low temperature experience (McLeod *et al.* 1985). In *Rhagoletis pomonella* the length of time before diapause termination occurs after exposure to higher temperatures decreases as the length of the cold period preceding morphogenesis increases (Neilson 1962; Laing and Heraty 1984). Experiments have shown that a prolonged exposure to cold is required for diapause termination in face flies, *Musca autumnalis* (Valder *et al.* 1969). In the field, face flies are exposed to low temperatures for 5 or 6 months, after which populations appear to emerge from diapause almost simultaneously (Matthew 1961) and with a full complement of fat, which is rapidly utilised for egg development (Hammer 1942). A laboratory study supported the field observations as females terminating diapause were found to have developed ovaries and deteriorating fat bodies (Caldwell and Wright 1978). Other examples of prolonged cold exposure requirements for diapause termination include 4 weeks at 4 °C for the chrysomelid *Plagiodera versicolora* (Stevens and McCauley 1989).

In some insects diapause termination is not affected by photoperiod at all. The reproductive diapause of the collembolans *Orchesella cincta* and *Tomocerus minor*, for example, terminates in response to temperature only (Woude and Verhoef 1988), diapause being ended more quickly at higher temperatures.

Similarly chilling is not a prerequisite for the completion of diapause in all insects (Danks 1987; Hodek and Hodková 1988; Zaslavsky 1988) and several examples of those where low temperatures play no direct role are available from the literature. These include the pentatomid bugs *Aelia acuminata* (Hodek 1975) and *Dolycoris baccarum* (Hodková *et al.* 1989). However, in many of these cases, although low temperatures may not be essential for the completion of diapause they can still be important in other ways, acting, for example, to slow the utilisation of energy reserves.

Interaction of photoperiod and temperature cues   The discussion of induction of the overwintering state indicated that insects frequently respond to the interaction of several stimuli rather than to a single cue. Such interactions have also been found to predominate in diapause maintenance and termination, although (as was seen above) a few cases where both photoperiod and temperature act alone to control diapause have been recorded. For example, diapause development is thought to proceed under short-day regimes without the need for exposure to low temperatures in the pentatomid *Plautia stali* (Kotaki and Yagi 1987). Most studies of temperature cues in diapause termination have concentrated on the interaction of temperature and photoperiod and, as in diapause induction, temperature has frequently been found to modify the reaction to photoperiod in diapause termination. Interaction of temperature, photoperiod (and other factors) can occur both simultaneously or serially.

A detailed study of diapause of the Colorado beetle, *Leptinotarsa decemlineata*, has been made at two sites in New York State (Tauber *et al.* 1988b). This study indicated that at the end of the summer, in September and October, diapause is maintained by photoperiod and temperature. However, by February beetles held at high temperatures of about 20 °C did not respond to photoperiod, whereas those held at low temperatures (15–18 °C) throughout dormancy were responsive to photoperiod. Intensity of diapause changed gradually as winter progressed, but sensitivity to photoperiod was maintained at low temperatures, even after emergence.

Temperature also modifies the photoperiodic responses in the firebug, *Pyrrhocoris apterus*. Temperatures of 15 °C or less prevent termination of diapause, which occurs when these insects are exposed to long days, whereas the response is not prevented if they experience higher temperatures (26 °C) (Hodková and Hodek 1987). Further, in this species the summation of photoperiodic signals was found to be temperature-dependent. Similarly, day-lengths, which at 5 °C result in diapause termination in the fly *Drosophila transversa*, do not have any effect at an ambient temperature of 0 °C (Kuznetsova and Tyshcherka 1979).

A study in Italy of diapause termination in the codling moth, *Cydia pomonella*, has indicated that under long-day conditions temperatures acted through the accumulation of day-degrees. However, the duration of diapause was also affected by the photoperiod at the time of induction, with shorter photophases resulting in longer diapause. Thus, the sum of effective temperatures required for adult emergence of 50 per cent of a diapause population differed according to the day-length experience before diapause (Deseö and Briolini 1986).

Many other examples of the interaction of factors in diapause maintenance and termination have been described, such as in the heteropteran *Cletus punctiger* (Ito 1988b), and several of these are highlighted in later sections of this chapter.

Changes in termination responses to temperature as diapause progresses
A further complication in the regulation of diapause termination by temperature arises from the observation that the responses of some insects to temperature differ at the various stages of diapause development. These changes can occur at different points in the progression of diapause, depending on species, and can occur either gradually or fairly abruptly.

A good example of such changes in temperature responses is provided by a study in Britain of the seed wasp, *Megastigmus spermotrophus* (Hussey 1955). During the first 15 weeks of diapause in this insect there is a decrease in both the length of time needed for emergence from diapause and in the number of larvae remaining in diapause when they experience warm conditions. This reflects a general downward change in their response to temperature, which is followed by a reversal of these changes as diapause progresses further.

Similar gradual changes in diapause regulatory responses have been recorded in a wide range of species, such as the dragonflies *Lestes sponsa* and *Aeshna mixta*, the moth *Lymantria dispar* and in *Chesias legatella* (Corbet 1956; Masaki 1956; Schaller 1968; Wall 1974). Rapid changes in thermal responses have been described in such insects as the grasshopper,

Austroicetes cruciata (Birch 1942), where egg-hatch responses at high temperatures change suddenly just before the onset of winter.

Diapause responses to temperature as an adaptation to environment In constantly changing environments the modifying or controlling effects of temperature on diapause induction, maintenance or termination allow insects to take advantage of the unpredictable opportunities offered in later summer or early spring.

A good example of this is the carrot fly, *Psila rosae*, in Britain. Carrot fly is a pest of a variety of plants, including carrots, parsnips and parsley and wild plants such as hemlock. It usually has two complete generations per year and will overwinter as both larvae and pupae (Wright and Ashby 1946). Diapause is known to occur in overwintering pupae (Biernaux 1968; Brunel and Missonnier 1968; Jørgensen and Thygensen 1968; Burn and Coaker 1981). Temperatures inducing diapause vary from 8 °C early in the year to 15 °C later (Brunel and Missonnier 1968; Städler 1970) and the newly formed pupae are the most sensitive to these diapause-inducing conditions (Naton 1966). It has been suggested that the duration of the overwintering larval stage may determine whether pupae enter diapause because at high temperatures larval development is quicker and the proportion of pupae susceptible to diapause-inducing conditions is larger. Conversely, at low temperatures larval development takes longer and reduces the proportion reaching the susceptible early pupal stage (Burn and Coaker 1981). This mechanism could result in a variable proportion of larvae and pupae in overwintering populations and allow the development of a third generation of adult flies in a year if conditions in October and November are favourable (Coppock 1974; Burn and Coaker 1981). Further work may be needed to confirm this.

Non-diapause responses to temperature Reaction to temperature can enhance insect survival during winter even in the absence of diapause. The intertidal collembolan *Anurida maritima* is a univoltine species that overwinters in the egg stage. Most of the eggs laid during the summer nter a diapause that is terminated in autumn but, due to low temperaturesin winter, further egg development is suppressed until spring. This strategy allows the species, which normally forages during low tide on the open shore and seeks refuge underground before the incoming tide (Joose 1966), to survive the winter period when they are too sluggish to find sufficient food in the limited period of low water (Witteveen *et al.* 1988).

Postdiapause responses to diapause temperatures Temperature is frequently a primary factor determining both the rate at which the characteristics of diapause are lost and the rate of postdiapause development (Tauber and Tauber 1976b; Tauber *et al.* 1986).

The effect of the length of hibernation (cold treatment) on the development time of the embryos of eight species of tettigoniid from central Europe has been studied in the laboratory (Ingrisch 1985). Two different types of response were observed when insects were subjected to a variety of chilling periods ranging from 3 to 12 months. In the first, length of hibernation and hatching date were either not connected or connected only after 3 months of cold treatment. The second response resulted in the time required for postdiapause development being significantly reduced by each 1-month increase in the length of cold treatment up to a total of 6 months exposure. The results of these responses are such that in late-hatching species hibernation length has little effect and a high thermal threshold prevents premature hatching, whereas in early-hatching species the potential speed of postdiapause development increases with the length of experience of cold conditions. The second strategy was interpreted as a response to a necessity of hatching at low to moderate temperatures.

Other species of insect in which temperatures during diapause and duration of diapause affect postdiapause responses include the ichneumonid *Pleolophus basizonus* and the katydid *Ephippiger cruciger* (Dean and Hartley 1977; Griffiths 1969).

Moisture

As was the case for induction of diapause, most of the work on the effect of abiotic cues on the maintenance and termination of the overwintering state has concentrated on photoperiod and temperature. The few studies of moisture effects that have been made have often involved egg diapause and have usually met with the problem of separating the requirement of many insects to absorb water during normal development from water absorption as a stimulus controlling diapause (see p. 38).

Despite this problem a few experiments have yielded strong evidence of moisture influencing duration and termination of hibernal diapause. For example, a study of the effect of drought on several species of European Tettigoniidae (Ingrisch 1986) has revealed that between one and seven cold treatments are needed to enable eggs to complete initial diapause if drought stress was not experienced. In Central European species the number of eggs maintaining initial diapause increased significantly when the eggs had no contact with water at the time when diapause should end. In contrast, the number terminating initial diapause in one species from Greece – *Tettigonia caudata* – which had experienced a period of drought, was highest in those eggs that had lost most water. The time during which moisture levels are important in diapause can be very short in some species. The larvae of the midges *Contarinia tritici* and *Sitodiplosis mosellana* respond to low soil moisture levels at the end of winter by remaining in diapause for an extra year. However, they only respond to soil moisture during a short (6-week) period (Basedow 1977). High moisture levels during this period result in diapause ending.

The precise definition of diapause termination can determine whether moisture is viewed as affecting diapause or postdiapause development in insects such as the European corn borer, *Ostrinia nubilalis*. This species relies primarily on photoperiod to control diapause, but post-diapause morphogenesis cannot proceed without water uptake (Beck 1967, 1980). Another example involves diapause maintenance in the Colorado beetle, *Leptinotarsa decemlineata*, which is controlled by the interaction of photoperiod and temperature (see p. 52), but after comple-tion of diapause the beetles remain in a quiescent state maintained by low temperature, low humidity or lack of food (Lefevere and DeKort 1989).

In most of the insects that have been studied, similar problems of determining whether moisture influences diapause termination or post-diapause development are encountered (e.g. *Aedes vexans*, *Contarinia sorghicola* and *Euxoa sibirica*) and in some cases the picture is further complicated by possible influences of other factors of the environment, which affect egg-hatch or development (Wilson and Horsfall 1970; Horsfall *et al.* 1973; Oku 1982; Baxendale and Teetes 1983). However, Tauber *et al.* (1986) point out that irrespective of which part of the life cycle moisture affects, the ecological consequences will frequently be similar, as development will be delayed until moisture conditions are favourable.

Biotic cues

Crowding

On page 39 it was concluded that although crowding can be an import-ant influence on the induction of summer diapause (e.g. the sycamore aphid, *Drepanosiphon platanoidis*) and diapause in stored product insects (e.g. the Indian-meal moth, *Plodia interpunctella*) it is rarely considered as an important factor in the production of the true over-wintering state.

Crowding is also frequently linked with the maintenance or termination of aestival diapause and diapause in store product insects. For example, the length of summer diapause in the sycamore aphid, *Drepanosiphon platanoidis*, is affected by the degree of crowding experienced not only by the diapausing aphids but also by their mother (Dixon 1975; Chambers 1982). Diapause termination is also promoted by crowding in a dermestid, *Trogoderma variabile*, by a high larval density (Elbert 1979).

However, few examples of crowding clearly influencing termination or maintenance of the true overwintering state are available in the literature and, once again, it cannot be considered an important factor.

Nutrition

As discussed previously (see p. 40), some insects make use of the changes in quality or quantity of the food available through the year to either trigger the induction of an overwintering state or modify the effect of other cues that induce overwintering. Both the quality and quantity of nutrition available to insects has also been linked to termination of diapause. As is the case for diapause induction, this is more frequently associated with aestival diapause but, in a few cases, both quality and quantity of food has been shown to affect the termination of the overwintering state.

The availability of prey has been found to be one of the factors important in ending hibernal diapause of predatory species such as the antlion, *Myrmeleon formicarius* (Furunishi and Masaki 1981). Reproductive diapause of the lacewing, *Chrysopa carnea*, ends when long spring days occur, coupled with the re-appearance of prey species (Tauber and Tauber 1973d, 1982) and in the dipteran *Chaoborus americanus* long days and availability of prey are the two factors necessary for the termination of hibernal diapause (Bradshaw 1970).

Quality of nutrition is also important in some species. Overwintering females of the Sitka spruce weevil, *Pissodes strobi*, undergo a reproductive diapause that usually occurs as a reproductive delay (Chapman 1971) with oocytes failing to mature until diapause is terminated. Temperatures and photoperiod cues have been shown to be necessary for ovary maturation (Harman and Kulman 1966) but females kept under identical conditions of temperature and long-day photoperiod, mature their ovaries if fed on Sitka spruce terminals but not if fed on lateral branches (Gara and Wood 1989).

The duration of diapause in *Delia brassicae* and the percentage of pupae in diapause may be affected by the day-length, temperature and host plant quality experienced by the adult flies and larvae (Hughes 1960; Read 1969). In the Indian-meal moth, *Plodia interpunctella*, the quality of the available nutrition appears to be important and the addition of an extract of rice bran to the diet influences termination of hibernal diapause (Tsuji 1966). However, the mechanism by which this response occurs has not been determined.

Behaviour leading to successful overwintering of insects has also been shown to be influenced by nutritional factors. For example, in a study of *Cletus punctiger*, Ito (1988a) showed that the hiding behaviour of diapausing adults in hibernacula is triggered by the reduction of food supply after a period of feeding.

Diapause induction and the length of diapause

Stimuli acting on the insect during or at the end of diapause are not the only factors that have been implicated in determining the duration of diapause. Although in many insects day-length at the time of induction has no effect on the maintenance or termination of diapause, in others the conditions prevailing at this point have been shown to affect the duration of the diapause period.

The precise factor affecting diapause duration once again varies between species. In some the number of short days experienced at critical stages of development are important. If the tobacco hornworm, *Manduca sexta*, for example, is exposed to longer day-lengths as an embryo or in the early larval instars, followed by shorter day-length in later instars, then it tends to undergo a longer diapause than if exposed to short days in both early and later larval development (Denlinger and Bradfield 1981). It is thought that this avoids individuals that enter diapause early also emerging too early and being exposed to the harsh conditions at the end of winter. Changing day-lengths also affect the lacewing, *Chrysopa carnea*, where decreasing photoperiods during diapause induction again result in longer diapause (Tauber *et al.* 1970).

Diapause duration in other species can be affected by the relationship between the photoperiods experienced by the insect and the critical photoperiod. In the two-spot ladybird, *Adalia bipunctata*, diapause induced by photoperiods near the critical level is of shorter duration than if induced by shorter photoperiods (Obrycki *et al.* 1983). The midge *Aphidoletes aphidimyza* responds to day-lengths close to the critical photoperiod by entering a longer diapause than it does when exposed to shorter days (Havelka 1980).

The length of diapause of other insects is related more precisely to the day-length experienced during the sensitive stage. As has been described, the lacewing, *Chrysopa carnea*, undergoes a longer diapause if it experiences shortening photoperiods during the diapause induction phase. This is not, however, an all-or-nothing response as the length of diapause also varies with the precise length of the photoperiod experienced.

Environmental variability and extended diapause

In many insect populations extended dormancy can be caused by either premature entry into, or prolongation of, diapause, which, as has been seen, may be the result of density-dependent or density-independent factors (Hanski 1988; Hanski and Ståhls 1990).

Most theoretical models assume that the function of extended dormancy is risk spreading. The anthomyiid group, *Pegomya geniculata*, are fungivorous insects that are restricted to a few species of *Leccinum* (Hanski 1989) in which the production of fungal sporophores becomes increasingly variable with increasing latitude. This factor should increase the incidence of prolonged diapause among species prevalent at higher latitudes. A study in Finland has shown that all seven common species found in Lapland have prolonged diapause (Hanski and Ståhls 1990) compared to only one of 14 species found in southern Finland, thus supporting the predictions of the risk-spreading models.

Re-entry into a second period of winter diapause

The adults of a very small number of insect species have been shown to undergo more than one winter diapause. These tend to be insects with long-lived adult stages, such as the seven-spot ladybird, *Coccinella septempunctata*. After the first period of diapause this beetle undergoes a period of egg-laying before entering a second diapause (Hodek 1976; Hodek *et al.* 1977). A second diapause can also be induced after a long period of egg laying in the Colorado beetle, *Leptinotarsa decemlineata*, and the lacewing, *Chrysopa carnea* (De Wilde *et al.* 1959; Tauber and Tauber 1976b).

A re-entry into diapause which does not require oviposition first has been demonstrated in females of the predatory mite, *Amblyseius potentillae* (Van Houten 1989). The first diapause is induced by short-day photoperiods and diapause development proceeds slowly under short days but is accelerated by long days. Under short days, sensitivity to photoperiod diminishes gradually and ultimately disappears completely, but after completion of diapause reappears again. A second diapause can then be induced in postdiapause females, again in response to short-day photoperiods and completed under long days.

Other species that can re-enter diapause include the orthopteran *Oedipoda miniata*, the coleopterans *Aelia acuminata*, *Agonum assimile* and *Riportus clavatus*, and the hemipteran *Chilocorus bipustulatus* (Hodek 1971b; Pener and Broza 1971; Neudecker and Thiele 1974; Numata and Hidaka 1982; Zaslavsky 1984). With the exception of the grasshopper, *O. miniata*, these insects usually live for more than a year and the return of photoperiodic sensitivity is necessary for diapause induction to take place in successive years. It is possible that similar mechanisms occur in other long-lived insects that hibernate more than once (Hodeck 1979). Continuous sensitivity to photoperiod in *O. miniata* may be an adaptation to ensure optimal timing of oviposition period in different habitats (Pener and Orshan 1980).

The small number of insects that have been recorded as being able to enter a second period of diapause may be a reflection of the relatively few studies that have been made.

Sex and the response to overwintering cues

Several instances of different sexes of the same species of insect varying in their responses to the same overwintering cues have been observed. Such differences have been found in both induction and termination of diapause although most studies have centred on diapause induction.

Diapause in the Southwestern corn borer, *Diatraea grandiosella*, has been shown to be dependent on both photoperiod and temperature. High temperatures can override photoperiodic determination of diapause, temperature cycles influence photoperiodic determination, and high temperatures and long days accelerate diapause development (Chippendale and Reddy 1973; Chippendale *et al.* 1976). In the critical region of the photoperiodic response curve females showed a higher incidence of diapause than males. Females also enter diapause later than the males, but resume active development earlier (Takeda and Chippendale 1982). Sex-linkage is also involved in controlling the rate of diapause development.

The beech leaf mining weevil, *Rhynchaenus fagi*, provides a good example of differences between sexes in their response to diapause terminating cues. In this insect only the female responds to a long-day stimulus by terminating diapause (Bale 1979).

Differences between sexes in their response to diapause cues take a variety of forms. The critical day-length for diapause induction in the small fruit fly, *Drosophila triauraria*, does not differ between males and females (Kimura 1988b), but male diapause is weaker than female diapause. The critical day-length and the diapause rate was found to vary geographically, and in crossing experiments the critical day-length and the diapause duration were found to be inherited in a quantitative manner. Inheritance of diapause ability has been studied in several other experiments involving crossing non-diapausing and diapausing strains of the same species. In a Polish study, for example, the diapausing trait of a strain of the two-spotted spider mite, *Tetranychus urticae*, exhibited incomplete dominance over the non-diapausing trait of two other strains (Ignatowicz 1986), and maternal effects were present in crosses of females of both non-diapausing strains with diapausing males. Two strains of *Drosophila funebris* from Japan enter a reproductive diapause in response to photoperiod and temperature but differ in their intensity of diapause. Crossing experiments suggest that diapause intensity may be controlled polygenically (Watabe 1988).

The seasonal morphs that are associated with diapause in butterflies are also frequently determined by environmental cues (Beck 1980; Saunders 1982). In Japan, for example, the pierid *Eurema hecabe* responds to photoperiod and decreasing temperatures in autumn by producing a distinctive autumn (diapause) morph (Yata 1974; Kato and Sano 1987). It was noticed that, in autumn, males of the summer morph remain more abundant than females of this morph (Kato 1986), and it was suggested that there may be a sex-linked difference in the photoperiodic response for seasonal morph determination. Laboratory experiments and field observations (Kato and Sano 1987) have shown that a sexual difference in the photoperiodic response does occur. Under short days at 25 °C the proportion of the autumn morph is higher in females than in males and as the rearing temperature is decreased the proportion of the autumn morph increases in both sexes. In this species sex-linked differences in the response for seasonal morph determination may provide the physiological basis for prediapause mating. Other examples of sexual differences in the production of seasonal morphs in butterflies include *Ascia monuste*, where such morphs are again regulated by photoperiod and temperature, but only in females (Pease 1962).

Crossing experiments using two subspecies of the swallowtail butterfly,

Papilio glaucus, with different diapause responses (Rockey *et al.* 1987), have indicated that the diapause responses of the hybrids were associated with sex and that the response may be determined by an X-linked gene. Studies of the genetics of photoperiodic control of the pupal diapause of two species of *Drosophila* have suggested that a single X-chromosome may control diapause in *D. lummei* (Lumme and Keränen 1978) and a single autosomal locus may do so in *D. littoralis* (Lumme and Pohjola 1980).

The various aspects of the winter environment also interact to affect insect survival and can act differently on males and females. For example, the ability of the Mediterranean fruit fly, *Ceratitis capitata*, to survive food and water deprivation is enhanced by a cold environment, which slows down the metabolism. The increase in survival time is, on average, greater in males than in females, although the most resistant females survive for the same length of time as the most resistant males (Carante and Lemaitre 1990).

Overwintering cues and parasitoid–host interactions

Parasitoids have a peculiar problem relating to diapause induction; to be successful the host and parasitoid life cycles must be well synchronised, as there is often a requirement for a host of a specific species and development stage. Accordingly, mechanisms of diapause induction, maintenance and termination must ensure careful tuning of parasitoid diapause to both host physiology and the environmental conditions that affect the host life cycle, as well as to the environmental conditions that affect the parasitoid itself. Three main modes of diapause regulation in parasitoids have been suggested (Tauber *et al.* 1983, 1986). Some show a total dependance on the hosts' physiological state, some are regulated by environmental cues similar to those affecting non-parasitoids, and others are regulated by both host physiology and environmental cues. Most species appear to respond to more than one cue (Saunders *et al.* 1970; Anderson and Kaya 1974, Eskafi and Legner 1974; Parrish and Davis 1978; Brodeur and McNeil 1989) and some authors suggest that inter-actions between cues are important (McNeil and Rabb 1973; Zaslavsky and Umarova 1981). As in non-parasitoid diapause, parasitoids usually show a well defined period during their development when they are sensitive to diapause-inducing cues, and a specific diapausing stage. Sensitive stages may overlap with the diapausing stage or they may be well separated, even in different generations.

The first mode of diapause regulation, dependence on the hosts' physiological state, has been demonstrated in several studies. Most authors studying this mode have concluded that some aspect of host physiology, often hormonal levels, affect maintenance or termination of diapause and play an important role in the synchronisation of parasite and host life cycles (Mellini 1983). Some of these hormonal cues are now well understood (Baronio and Sehnal 1980).

The tachinid *Pseudoperichaeta nigrolineata* is a parasitoid of the European corn borer, *Ostrinia nubilalis*. Parasitoid eggs are laid near the host larvae and, after hatching, the larvae penetrate the host. The host eventually dies 1 or 2 days before the fully grown parasitoid escapes. A close relationship has been described between some development phases of the parasitoid and events in the host life cycle such as moulting (Ramadhane *et al.* 1987). Ecdysteroid titres are very low during diapause in *O. nubilalis* (Gelman and Woods 1983) but rise before diapause termination (Bean and Beck 1983). When diapausing host larvae are parasitised *P. nigrolineata* larvae develop through the first instar in the same way as in non-diapausing host larvae but then stop growth in the second instar (Ramadhane *et al.* 1988), subsequently resuming growth only at host diapause termination. In the laboratory, injecting a high dose of ecdysterone has been found to induce both parasitoid growth and host apolysis, the parasitoid larvae increasing the host sensitivity to exogenous ecdysteroids (Ramadhane *et al.* 1988). The close synchronisation between parasitoid and larval development was broken by injecting a low dose of ecdysterone.

A similar situation has been described when tachinid parasitoids of the genus *Pseudoperichaeta* develop in the wax moth, *Galleria mellonella*. In this case modifications to the physiology of the host induce arrested development in the middle of the second instar in about one-third of parasitoid larvae (Grenier and Delobel 1984a), growth resuming some time later (Grenier and Delobel 1984b). It has been suggested that there are low ecdysteroid levels in hosts containing parasitoids in arrested development, and higher levels when parasitoid growth has been resumed (Plantevin *et al.* 1986).

Not all parasitoids merely react to the physiological changes in their hosts. As long ago as 1930 it was shown that many insect parasitoids are capable of altering the physiology of their hosts (e.g. Holdaway and Evans 1930; Varley and Butler 1933), and there are a few examples in the literature of regulation of the onset of host diapause by parasitoids. Non-parasitised larvae of the greenbottle, *Lucilia sericata*, have been shown to enter diapause in the final larval instar, whereas those that have been parasitised pupate first (Holdaway and Evans 1930).

A more recent study (Moore 1989) concentrated on a braconid wasp, *Cotesia koebelei*, parasitising the lepidopteran *Euphydryas editha*. Larvae of the butterfly normally enter diapause after passing through three feeding instars, although some individuals continue through a fourth instar. Parasitoid attack resulted in larvae being more likely to pass through the extra feeding instar, a response that may be invoked by a chemical injected at oviposition. It is suggested that delayed diapause may be necessary to allow the wasp sufficient time to complete prediapause development, or alternatively it may allow the parasitoid to synchronise its life cycle with an alternative host species, thus allowing it to complete an extra generation when *E. editha* is in diapause.

The environmental factors that influence the induction of the over-wintering state in parasitoids are similar to those used by non-parasitoids, and most investigations have shown them to be the most important regulatory cues. As in non-parasitoids, photoperiod has often been found to act as a cue for diapause induction. Several examples of responses to photoperiod are available in the literature (e.g. Askew 1971; Saunders 1982) and these include responses to changes in photoperiod (Schopf 1980). The cue has also been found to influence diapause maintenance (Claret 1973; Obrycki and Tauber 1979). An effect of both temperature and thermoperiod has been described in several species, sometimes as the sole cue (Flint 1980), but usually interacting with photoperiodic cues both for diapause induction (e.g. McNeil and Rabb 1973; Saunders 1973; Wylie 1980) and diapause maintenance (e.g. Wylie 1977; Nechols *et al.* 1980).

Less information is available on the effect of nutrition, crowding and moisture on parasitoid diapause. However, a few examples have been reported. A nutritional role for the host in diapause regulation in the chalcid wasp, *Melittobia chalybii*, has been deduced by Schmieder (1933). Either nutrition or parasitoid density appears to affect diapause induction (Schmieder 1939) in the ichneumonid wasp, *Sphecophaga burra*, and humidity appears to act as a cue for diapause induction and termination in at least three species of parasitoid (Eskafi and Legner 1974; Parrish and Davis 1978; Schopf 1980).

The third category of diapause regulation in parasitoids involves reliance on a combination of host physiology and environmental cues. The relative importance of the two types of cue has been shown to vary widely between species.

Mixed cue regulation is well illustrated by a study of *Aphidius nigripes*

(Brodeur and McNeil 1989), a braconid endoparasite of the potato aphid, *Macrosiphum euphorbiae*. In this species interactions between low temperatures and short day-lengths were found to be significant in inducing diapause, and sensitivity to these factors was greatest during embryonic and first instar development; maternal effects were also demonstrated. Females that experienced short day-lengths and low temperatures as adults produced more diapausing progeny, and maternal age also affected diapause in their offspring. Finally, a higher proportion of parasitoids developing in small, first instar aphids entered diapause than those in larger fourth instars. The interaction of such factors may be adaptations which allow the parasitoid to respond to seasonal changes in the environment more accurately.

Cataloaccus aeneoviridus, a parasite of the tobacco hornworm, responds to photoperiod, temperature, maternal age and the diapause status of the host for diapause induction (McNeil and Rabb 1973). A slightly different situation was highlighted by a study of the chalcid wasp, *Mesocomys pulchriceps*, which responds to photoperiod for diapause induction and interactions with the host for diapause maintenance (Van den Berg 1971). Several other examples of mixed cue regulation have been reported in the literature (e.g. Maslennikova 1958; Saunders 1965).

Overwintering cues for social insects

Social insects may overwinter either as solitary fertilised queens, which found annual colonies (frequently a feature of more primitive groups) or as colonies, either with or without food. Few experimental studies have been made of regulation of overwintering in social insects because of the difficulties of working with whole colonies but the information that is available, much of which is based on observational work, suggest that similar environmental cues to those that regulate overwintering in non-social insects are involved. In addition, there is interaction between colony members and the state of environmental stimuli; some colonies can achieve a degree of thermoregulation within the nest and in some the influence of the queen on worker behaviour towards larvae is important (Brian and Kelly 1967; Pener 1974; Seeley and Heinrich 1981).

Overwintering has been studied in two genera of ants, *Formica* and *Myrmica*. More work has been done on the latter genus, particularly on the red ant, *Myrmica rubra*, but the results concerning induction of the overwintering state are not clear. The experiments of Kipyatkov (1974, 1976) suggest that diapause in workers is induced by short day-lengths, that workers can communicate information on photoperiod and third instar larvae enter diapause in response to diapause induction in the workers. Queen diapause results from the combined cues of diapause induction in workers and a direct effect of photoperiod on the queen. However, other studies have failed to demonstrate any effect of photoperiod on diapause induction (Brian and Kelly 1967; Hand 1983). Instead, low temperatures and reduced food appeared to enhance diapause induction in gyne-biased third instar larvae. Worker-biased larvae were found to overwinter, but without diapausing, and it was concluded that this was the result of underfeeding. Tauber *et al.* (1986) have suggested that the differences between these two studies concerning the influence of photoperiod, may be the result of a seasonal variation in sensitivity of *M. rubra* to photoperiod (Kipyatkov 1979) and that a further study is required to investigate this possibility.

Among other more advanced insects that overwinter in colonies is the honey bee, *Apis mellifera*. Preparation for winter in this species normally involves development of food stores, cessation of reproduction and behavioural changes that reduce colony development. With the exception of the preparation of food stores, responses to decreasing photoperiods have been shown to bring about such changes through a reduced oviposition rate by queens, changes in worker behaviour and reduced feeding (Kefuss 1978). Colonies of *A. mellifera* can also achieve thermoregulation in nests during cold periods. Colonies form clusters in response to low temperatures occurring in the first part of the overwintering period and maintain the temperature above 9 °C at the edge of these clusters. Later in the overwintering period, the temperature of the whole hive is maintained at 30 °C or above (Seeley and Heinrich 1981). Stored food (honey) is used in the generation of the heat.

Primitive social insects, such as some species of social wasp (*Polistes*) overwinter as queens rather than as colonies and respond in both immature and adult forms to such factors as photoperiod, temperature and nutrition to signal the onset of winter (Bohm 1972; Brian 1979). In relatively advanced groups other factors peculiar to social insects are also important in the regulation of diapause induction, including the queen's age and the production of queen pheromones. The latter may in turn be influenced by photoperiod. The combination of these factors result in the production of new gynes that enter diapause and overwinter (see Tauber *et al.* 1986). In some species, which overwinter as fertilised females, male production may also be stimulated by photoperiod in autumn (Suzuki 1981).

Varying responses to overwintering cues in different geographical areas

Different individuals of the same species have often been shown to vary in their response to an environmental cue. Such variations occur in a number of ways. For example, they can be between individuals of different sexes (see p. 60) or from different geographical areas.

The optimum strategy for most insects is to maximise their reproductive potential during summer, but to do so in such a way that they do not reduce the probability of their offspring surviving winter. As unfavourable conditions approach, reproduction should not cease until there is just sufficient time for the insect to produce the stage that can survive these conditions. Species that have a broad geographical range may therefore be expected to modify their reactions to local environmental cues that indicate the onset of winter, and several authors have suggested that insects that have a wide range adapt their life histories to suit local conditions (Masaki 1978; Tauber and Tauber 1981; Bradshaw and Holzapfel 1983; Hodek 1983; Riedl 1983). Three factors have been found to be of particular importance: latitude, altitude and proximity to the centre of a large land-mass.

Latitude

The photoperiodic induction of diapause in the peacock butterfly, *Inachis io*, varies between populations and appears to be finely tuned to local conditions (Pullin 1986). In southern Britain, where the species is usually univoltine, the photoperiodic threshold appears to correspond with the longest day-length experienced at the latitude at which the butterflies are found. The critical day-length in France is shorter and allows two generations per year. Thus, the peacock butterfly balances the risk of mortality caused by an early winter for which it is unprepared, against the need for each population to maximise its reproductive potential, taking account of the different local conditions experienced at different latitudes.

Several other examples of differing responses at different latitudes to token stimuli indicating forthcoming harsh conditions have been reported in the literature.

The bertha armyworm, *Mamestra configurata*, overwinters in the pupal stage in western Canada and the induction of diapause in pupae is dependent on photoperiod and temperature (Hegdekar 1977). As both day-length and its rate of change are a function of latitude, Hedgekar (1983) has studied the effect of latitude on the critical photoperiod for diapause induction and found that there is also a strong relationship between critical photoperiod and higher latitude. An increase in 1°14′ in latitude was associated with an increase of 22 minutes in the critical photoperiod.

In both Europe and North America the codling moth, *Cydia pomonella*, enters diapause well in advance of the onset of harsh winter conditions. The initiation of diapause is caused by a combination of factors such as shorter photoperiods, lower temperatures and poorer food quality, but photoperiod has been shown to have the major role (Peterson and Hamner 1968). The photoperiodic reaction shows a similar variation in both Europe and North America (Riedl and Croft 1978). An increase of 10° in latitude corresponded to an increase of 75 minutes in critical photoperiod. Further, the critical photoperiod varied between years for individuals from the same geographic populations and it was suggested that this was probably due to the modifying effect of prediapause temperatures.

The response of some aphids which use the length of the night as a seasonal cue to time the production of sexual morphs also varies with latitude (Shaposhnikov 1981).

Altitude

A clear demonstration of differences between insects at different latitudes and altitudes is found in diapause induction in *Wyeomyia smithii* (Bradshaw 1976). Short days induce the overwintering state of this mosquito (Smith and Brust 1971; Bradshaw and Lounibos 1972) but critical photoperiods are not the same in populations from all areas of its range (Fig. 3.10a). Specimens from higher latitudes and altitudes respond to longer day-lengths than those from more southerly areas or lower altitudes. This variance can be explained by the length of the growing season of the mosquito's breeding site, the pitcher plant, which is shorter in more northerly latitudes (see p. 40) or higher altitudes. The adaptation to photoperiod enables the mosquito to synchronise its life cycle with the length of this growing season in its own area (Fig. 3.10b).

Geographical area

The third factor that has been found to be important is the proximity to the centre of a land-mass. The burnet moth, *Zygaena trifolii*, shows differences in its responses to photoperiod at different latitudes and distances from the sea (Wipking and Neumann 1986). In this study,

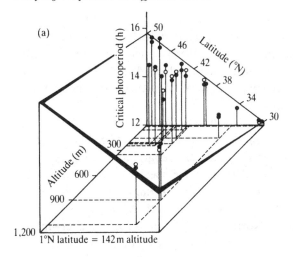

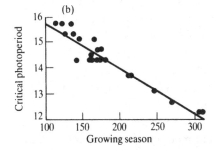

Fig. 3.10. (a) The relationship between critical photoperiod for the induction of overwintering state in *W. smithii*, latitude and altitude. The variables are related to the equation:

$$\text{critical photoperiod} = 6.51 + 0.185 \text{ latitude} + 0.00129 \text{ altitude}.$$

(b) The relationship between critical photoperiod for the induction of the overwintering state in *W. smithii* and the length of the growing season of the pitcher plant (reprinted with permission from *Nature* (Bradshaw, 1976) © (1976) Macmillan Magazines Limited).

rearing the moths at various photoperiods between LD 8:16 and 16:8 and at 20 °C, showed that insects from a Spanish population had a critical photoperiod of 14.5 hours for diapause induction, whereas all the larvae from a West German population is univoltine unless the generation time is prolonged to 2 or 3 years by extended diapause or by repetitive diapause reactions.

An ability to vary the reaction to diapause-inducing cues can be an important determinant of how far an insect can extend its range, particularly if the extension involves moving from the tropics to a temperate area. In a study of nine species of flesh flies of the genus *Boettcherisca*, Kurahashi and Ohtaki (1989) concluded that different species have evolved their pupal diapause independently of one another as an ecophysiological mechanism that enabled them to expand from tropical to temperate regions.

The effect of the geographical position of an insect population is not solely confined to the induction of the overwintering state. The progression and termination of diapause in different geographical areas has also been considered. In Japan, diapause in three species of small fruit flies, *Drosophila auraria*, *D. triauraria* and *D. subauraria*, is induced by short day-lengths but was found to be deeper in Northern populations. This variation may occur because Northern populations enter the overwintering state earlier and experience relatively long day-lengths, which are less effective in maintaining diapause. A fourth species, *D. biauraria*, showed a deep diapause when compared with the other three species, but did not show clonal variation in diapause intensity (Kimura 1988a).

The seasonal progression and termination of diapause in the Colorado beetle, *Leptinotarsa decemlineata*, which was described on page 52, was studied at two sites in New York state, one at Riverhead (40°58'N) and one at a site with a cooler climate and heavier soil at Freeville (42°21'N). Overwintering populations in both areas were composed of adults from both the first and second summer generations (Tauber *et al.* 1988a). The seasonal progression of diapause in the two populations was shown to be similar and both had similar temperature thresholds for reproductive development. However, the diapause in the warmer Riverhead population was less intense (shorter) and a lower thermal time was required for initiating oviposition. It was suggested that this may be an adaptation to a possible earlier spring growth of the beetle's host plants in the lighter soils and warmer climate of Riverhead (Tauber *et al.* 1988b).

Variation in response to overwintering cues between different individuals from the same geographical area have also been reported. There are, for example, at least two strains for the European corn borer, *Ostrinia nubilalis*, in Quebec with responses that allow one strain to complete only one generation per year while the other can develop a second. Two attributes of the latter strain promote the development of a second generation: first, it emerges from overwintering earlier in spring and generally shows a less intense diapause; and secondly, it has a lower critical photoperiod for diapause induction (McLeod *et al.* 1979).

Synchronisation of overwintering stage with season

Clearly it is important that insects pass the winter in the developmental stage that maximises their chances of surviving harsh conditions (see Chapter 5) and environmental stimuli such as photoperiods can be used to control the production of morphs or the rate of development of populations to achieve this goal. In diapause, environmental stimuli triggering the production of the overwintering stage can be different from those triggering diapause induction (e.g. different photoperiods acting at different stages of the life cycle). Further, as has been seen, populations from different geographical areas show different responses to the same environmental stimuli to adapt themselves to their own particular habitat (see p. 67).

For many insects, as winter approaches the optimal strategy is to cease reproduction before entering a quiescent overwintering stage, as late as possible (see p. 67). However, the overwintering site must be found in time to allow the physiological and behavioural adaptations that are needed to survive winter (Tauber *et al.* 1986).

Aphids such as the bird cherry-oat aphid, *Rhopalosiphum padi*, which have life cycles similar to that described on page 27 illustrate this well. This species feeds on various grasses during the summer and responds to short day length and low temperatures in autumn by producing special winged morphs, gynoparae, which fly to their overwintering host – bird cherry, *Prunus padus*. Upon reaching this plant the gynoparae give rise to the sexual morphs (oviparae), which produce the overwintering eggs. The gynoparae should ideally be produced when the length of time before leaf fall is equal to the time needed for migration, development of oviparae, mating and oviposition (Dixon 1985, 1987). Thus gynoparae of *R. padi* are produced late in the season and only have a short period of time between arriving on the winter host and leaf fall, in which to produce offspring that in turn will lay eggs (Dixon 1976). Two strategies could now be followed. If the aphids feed, thus prolonging adult life, and give birth to offspring throughout adult life (a reproductive schedule similar to the summer morphs), then a large proportion of the offspring will have insufficient time before leaf fall in which to mature and lay eggs (Fig. 3.11). Instead, a second strategy is adopted. The gynoparae do not feed on the overwintering host, but complete their reproduction within a few days of landing on the plant and die (Fig. 3.11), thus maximising the chances of their offspring developing to the adult stage and laying eggs before being killed by leaf fall (Walters *et al.* 1984). There is some evidence

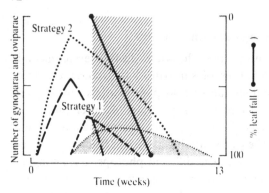

Fig. 3.11. Graphical representation of the consequences for the same number of gynoparae that, after colonizing their primary host, follow either a typical gynoparae reproductive strategy (strategy 1; ——— number of gynoparae, ——— number of oviparae) or an alate exule (strategy 2; · · · number of gynoparae, · · · number of oviparae). With the same reproductive strategy as alates exules (strategy 2) leaf fall (striped fields) would cause the early death of many of the adults and loss of their offspring and in the absence of leaves they would be unable to complete their reproductive schedule (dotted fields) (after Bonnemaison 1951; Dixon 1976).

to suggest that if the aphids land on less mature leaves, which will remain on the plants for longer, they may feed, prolong their lives and increase their rate of reproduction (Walters, unpublished data). In this example the aphids react to a seasonally changing environment in such a way that they fly to their overwintering site in just enough time to produce the overwintering morph.

The black been aphid, *Aphis fabae*, has a similar life cycle to *R. padi* but lays overwintering eggs on various species of spindle, *Euonymus*. Field studies of this species of aphid have revealed that in some years heavy infestations of the leaves of the overwintering host are still present at the time the leaves abscise, and this results in many aphids failing to lay eggs. Thus, in the black bean aphid (as in *R. padi*) the time of the return migration in relation to the state of senescence of the leaves of the winter host has an important effect on the numbers of eggs laid (Saynor and Seal 1976).

Whereas aphids such as *R. padi* move to their overwintering host when there is just enough time to produce the overwintering stage of the life cycles, the birch aphid, *Euceraphis punctipennis*, lives on birch throughout the year, laying overwintering eggs in autumn. Unlike *R. padi* and *A. fabae*, this species can feed on small twigs as well as leaves (Stroyan 1977) and thus autumn leaf fall does not signal the immediate end of its food supply. In mild autumns birch aphids have been found to prolong egg laying later into the autumn (Gange 1989), thus following a similar strategy of continuing reproduction as late into the autumn as possible, to that already described for *R. padi*.

Other insects use photoperiodic stimuli to adjust their rate of development to ensure that the correct stage is reached at the correct time of year. The rate of nymphal development of some crickets has been shown to be reduced by long days and increased by short days, and this is thought to ensure that the adult stage is present at the time of year that the overwintering eggs have to be laid (Masaki 1978).

Some parasitoids that overwinter within a host species have to adapt their life cycles to ensure that the host is parasitised prior to the onset of winter. *Meteorus trachynotus* is a bivoltine braconid parasitoid that requires two host species to complete its yearly life cycle. During the summer it parasitises several Lepidoptera but shows a preference for the spruce budworm when it is available (Krombein *et al.* 1979; Pogue 1985). It overwinters as an egg or larva in a host other than spruce budworm (Blais 1960) and probably in the tortricid *Choristoneura rosaceana* (Maltais *et al.* 1989). Diapause in the winter host *C. rosaceana* is induced by short day-length and low temperature experiences during the first and second instars (Gangavalli and AliNiazee 1985) and, as the parasitoid probably does not attack larvae after they have spun a hibernaculum, the temporal coincidence of prediapause host larvae and searching parasitoid adults is critical. The yearly life cycle of *M. trachynotus* is adapted to the environment so that females are searching for hosts while the host larvae are feeding and developing to the second and third instar stage, at which point they enter diapause (Maltais *et al.* 1989).

The bertha armyworm, *Mamestra configurata*, prepares for winter by tunnelling into the soil and making an earthen cell in which it normally pupates. This insect is parasitised by an ichneumonid wasp, *Banchus flavescens*, and this parasitoid adjusts the duration of its early instars to allow the host to complete its larval development and tunnel into the ground before it dies (Arthur and Mason 1985). If a first instar host is parasitised then the first two instars of *B. flavescens* can last for up to 22 days, whereas if a third instar host is parasitised, the first four instars of the parasitoid can complete development in approximately the same number of days. After the host's earthen cell has been made, the parasitoid develops rapidly and the fifth instar larva kills and leaves the host and forms a cocoon within the cell, where it remains during the winter.

These examples again illustrate the problem faced by some parasitoids that was highlighted on page 62. As winter approaches they not only have to prepare for the survival of harsh conditions themselves but also have to be adapted to changes which prepare their hosts for winter.

In some insects the exact stage arrived at before winter is not quite so critical in determining survival. The red turnip beetle, *Entomoscelis americana*, overwinters in the egg stage. Eggs are laid on or near the surface of the soil. Embryo development begins soon after laying and continues until a fully-developed embryo is formed. The mature embryo then enters diapause, from which it emerges in March or April (Gerber and Lamb 1982). It has been shown that eggs laid late in the year do not reach the mature embryo stage and enter diapause before the onset of the cold weather. However, provided the embryos have reached a critical minimum stage of development (germ band formation) they can still overwinter successfully. In this way the beetle is able to cope with any particularly short summer seasons, which may reduce the rate of their growth and development. There is also a second advantage resulting from this overwintering strategy. As they survive winter in a wide range of embryonic stages, egg-hatch the following year is staggered and this enables *E. americana* to utilise a larger variety of food plants, thus avoiding catastrophes resulting from poor synchronisation between hatching and early season plant growth (Gerber 1984). Another example arises from a study in Quebec, in which the univoltine lepidopteran *Ctenucha virginica* was shown to be able to overwinter in any of three late larval instars. This ability was thought to help the insect to synchronise its diapause initiation with the onset of low temperatures, which was especially important in this species as larvae died when kept at high temperatures and long day-length during diapause (Fields and McNeil 1988).

Emergence from the winter state must coincide with the onset of more favourable conditions and many examples of adaptations to achieve this are known (see p. 42). Adult coccinellids of the species *Hippodamia convergens* form large overwintering aggregations in the Sierra Nevada mountains of California. The maintenance and structure of these aggregations, which can remain together for up to 10 months (Hagen 1962), is helped by the release of an attractant pheromone (Copp 1983). The extended dormancy period functions to synchronise the feeding and reproductive activities of this univoltine species with their aphid food supply (Bennett and Lee 1989).

4

Insect cold-hardiness

Introduction

Cold-hardiness is the most well documented aspect of insect over-wintering, with a number of comprehensive reviews (Salt 1961; Danks 1978; Zachariassen 1985; Cannon and Block 1988; Bale 1989) and many shorter review articles on limited parts of the subject including super-cooling (Sømme 1982), ice nucleating agents (Zachariassen 1980, 1982; Duman 1982; Duman and Horwath 1983), protein antifreezes (Duman 1982; Duman *et al.* 1982; Duman and Horwath 1983), polyhydroxy alcohols (polyols) (Ring 1980; Zachariassen 1980), biochemistry and low temperature metabolism (Baust 1981; Storey 1984) and environmental triggers for cold-hardening (Baust 1982).

When Salt (1961) reviewed the 'principles of insect cold-hardiness', three main components were recognised: (i) cold acclimation; (ii) freeze tolerance; and (iii) freeze avoidance by supercooling. Over the last 25 years the subject has evolved to provide a clearer understanding of the structure and function of cryoprotective polyols and ice nucleating agents and thermal hysteresis (antifreeze) proteins have been discovered. This in turn has led to an expanding nomenclature in insect cold-hardiness where certain terms have often been used synonymously, e.g. supercooling and cold-hardiness, without clarification or definition, while other synony-mous terminologies (e.g. frost susceptible, freezing susceptible, freezing sensitive and freezing intolerant) have been used preferentially by different authors leading to confusion and inconsistencies in published reports. Increasingly there has been a recognition that a standard terminology is required and at the outset this chapter provides a list of definitions and concepts from the modern literature based mainly on terms described by Zachariassen (1985).

Much of this chapter describes the physiological and biochemical bases for the commonly accepted strategies of cold-hardiness, *freeze tolerance* and *freezing intolerance* (*freeze avoidance by supercooling*). However, Danks (1978) concluded that our understanding of insect overwintering was hindered by the disjunction between physiology/biochemistry and ecology, and noted particularly that most work on cold-hardiness was based on laboratory temperature studies rather than factors that affected the temperatures experienced by insects in the field. Bale (1987) has continued this theme and commented that although a number of insects have been studied independently for cold-hardiness characteristics and winter field survival, very few species have been investigated throughout the year in the combined ecophysiological laboratory and field approach commended by Danks. The benefits of this integrated type of study are described in this chapter.

Finally, two recent reviews (Baust and Rojas 1985; Bale 1987) have argued for a 'critical assessment' of the dogma surrounding the 'founding hypotheses' of insect cold-hardiness, particularly the effect of cooling and warming rates on the designation of a species as tolerant and intolerant of freezing (Baust and Rojas 1985), and the possible occurrence of 'prefreeze' mortality in insects regarded as freezing intolerant (Bale 1987) and these views are also considered in this chapter.

Concepts and definitions

Cold-hardiness is the ability of an organism to survive at low temperature. Cold-hardiness varies between and within species and can be measured by a number of indices such as supercooling points or LT_{50}. The LT_{50} is the time (at a constant temperature) or temperature (for a fixed exposure period) required to kill 50 per cent of the population. Most research has been conducted on species from climates that are sporadically or continually below freezing (0 °C) in winter.

Freeze tolerance is the ability to tolerate the formation of ice in the body tissues and fluids. It is generally accepted that this freezing occurs extracellularly to prevent intracellular freezing, which would be lethal.

Vitrification (of aqueous solutions) is the process by which a supercooled solution becomes amorphous (glass-like) solid water. Partial vitrification, which may occur in insects, describes a mixture of ice and amorphous solid water.

Freezing intolerance (synonymous with *freezing susceptibility, frost susceptibility* and *freezing sensitivity*) describes species that are unable to tolerate the formation of ice in the body tissues and fluids.

Freeze avoidance is the strategy by which freezing intolerant insects survive at low temperature, based on their ability to *supercool*.

Supercooling is the phenomenon by which water and aqueous solutions remain unfrozen below their *melting point*. (The melting point is the temperature at which the last ice crystal disappears when a frozen sample is warmed.) The *supercooling point* is the temperature at which spontaneous freezing occurs in a supercooled liquid (also referred to as the *crystallisation temperature*).

Thermal hysteresis describes the phenomenon by which the freezing point of the haemolymph (*hysteresis freezing point*) is depressed relative to the melting point. Thermal hysteresis agents, which inhibit ice growth, are proteinaceous and are now more usually described as *antifreeze proteins*.

Cold acclimation (*cold acclimatisation* or *cold-hardening*) is the seasonal increase in cold-hardiness that occurs in most species from summer to winter so as to avoid cryoinjury or death at temperatures below which normal growth and development are possible.

Strategies of insect cold-hardiness

In principle, two main strategies of cold-hardiness have evolved in insects that are exposed to temperatures below the freezing point of their body fluids, as set out in Figure 4.1 (Baust 1982). Insects are either freeze tolerant or intolerant depending on their ability to survive the formation of extracellular ice (Salt 1961). Freeze-tolerant species often contain ice nucleating agents (proteins or peptides), which are normally only present in winter and ensure protective extracellular freezing at high subzero temperatures (Zachariassen and Hammel 1976; Zachariassen 1980, 1982; Duman 1980, 1982) and polyhydroxy alcohols (polyols), which function to limit freeze damage (Baust 1973; Duman and Horwath 1983). Freezing is lethal to freezing intolerant species and is avoided by supercooling – the body tissues and fluids are maintained in the liquid state below their equilibrium freezing point (Salt 1936). Seasonal increases occur in the concentration of one or more polyols or sugars, thus extending the inherent ability to supercool (Baust 1981; Sømme 1982), and the activity of antifreeze (thermal hysteresis) proteins (Duman 1977a,b), which lower the freezing point of the haemolymph relative to its melting point and may act to stabilise the supercooled state (Zachariassen and Husby 1982).

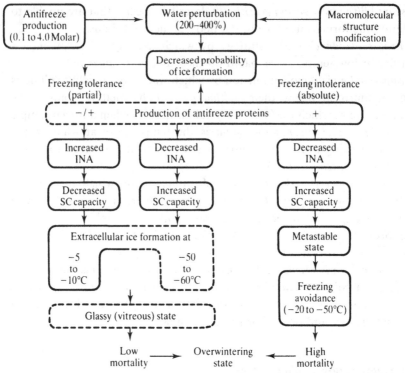

Fig. 4.1. Schematic representation of the principal adaptive strategies utilized by varying classes of winter hardy insects. INA, ice nucleating agents; SC, supercooling; ⬚ refers to features so far observed in a small number of species but which may be common in overwintering insects of the types shown (Baust and Hansen, personal communication).

Similar cryoprotective antifreezes and antifreeze proteins are found in both freeze tolerant and freeze intolerant species, although the roles of these compounds differ in the two types of insect. The characteristic difference between the strategies is the winter loss or masking of the nucleators in supercooling-dependent species and the synthesis or unmasking of nucleating agents in freeze tolerant species (Duman 1982; Baust and Rojas 1985).

Freeze tolerance

The majority of insects living in cool and colder climates are freezing intolerant; freeze tolerant species are usually found in regions where the winters are extremely cold, such as Alaska and Scandinavia. Many freeze tolerant species overwinter as larvae or pupae but freeze tolerant adults are not uncommon (Ring 1980). On balance, freeze tolerance appears to be more advantageous than freeze avoidance but it is not a risk-free strategy and mortality increases with decreasing temperature and increasing time in the frozen state (Sømme 1982; Zachariassen 1985). The designation of a species as freeze tolerant (or intolerant) is usually based on laboratory experiments conducted at a single and uniform rate of cooling and in a single exposure, in which the specimen is thawed after reaching its supercooling point. By contrast, in nature, cooling and warming rates are variable and the frequency of freeze–thaw cycles and the severity of frosts will differ in different regions and overwintering sites. Thus, while the success of the tolerance strategy is evidenced by the distribution of such species in the coldest parts of the world, where only a small number of freezing intolerant species with abnormally low supercooling points (and high cryoprotectant antifreeze concentrations) can survive, the 'cost' to freeze tolerant species in terms of the numbers of individuals that are killed in winter by freezing (and other cold-related processes) is largely unknown. This arises because, in common with freeze avoiding species, research has concentrated on the laboratory elucidation of biochemical and physiological mechanisms at the species level in isolation from ecological studies on the survival of natural populations.

Death by freezing

While it has long been recognised that freezing is a major threat to overwintering insects, the physiological bases of 'death by freezing' have been the subject of much speculation. The possible mechanisms of freezing injury in freeze tolerant insects have been described by Salt (1961), Baust (1973), Danks (1978), Block (1982a) and Zachariassen (1985) and are summarised below.

The stresses and strains of ice formation in living systems distort tissues and it has been proposed that this may lead to the mechanical damage of cells. However, there is no clear evidence for mechanical pressure and, furthermore, it has been argued that the lipoprotein network of the cell membrane would inhibit ice growth and that ice penetration of the cell was unlikely because of the very slow rate of ice growth permitted by the minute pores of the cell membrane. These views encapsulate the so-called 'site of freezing' theory, in which survival of freezing is achieved by the restriction of ice formation to extracellular areas and intracellular freezing is regarded as fatal; although Salt (1962) described intracellular freezing tolerance in insect fat body cells, there are very few reports of survival after intracellular freezing.

The theory of *electrolyte imbalance* relates to the increased solute concentration and dehydration that follows ice formation. When water freezes out of an extracellular aqueous solution, electrolyte levels are thought to increase proportionally and to eventually become lethal (Lovelock 1953). At the same time as the electrolyte concentrations increase extracellularly, intracellular water diffuses out in response to the changing osmotic gradient. Proteins may be denatured by the loss of bound water from their surface or denaturation may occur after cellular dehydration, that leads to the formation of abnormal disulphide bonds (Levitt 1962).

Two other theories mentioned by Baust (1973) – the *critical cell volume* and phenomenon of *recrystallisation* – are based on the ideas of Meryman (1970) and Luyet (1960), respectively. As the freezing process proceeds, cell water moves to the growing ice masses in the extracellular spaces, causing the cells to shrink. Loss of cell water is tolerable until a minimum size is reached (the critical volume) and shrinkage below this level would cause the membrane to rupture. Finally, it has been suggested that some insects may be able to survive the initial freezing even if the temperature continues to decrease. However, recrystallisation may occur either if the organism has to endure prolonged exposure or during a phase of warming before thawing. Generally, larger ice crystals grow at the expense of smaller ones and may result in osmotic stress (due to the rapidly elevating ice and electrolyte content and decreased water content) as well as mechanical damage.

In a balanced appraisal of these theories Baust (1973) concluded that a single site of freezing was unrealistic and that the freezing process and the associated deleterious effects represented a continuum in which individual events were largely inseparable. Equally important, a multifunctional role was suggested for glycerol, such that this polyol alone or in combination with others, could prevent or minimise all the envisaged freezing injuries or death. These mechanisms of cryoprotection in freezing tolerant insects are discussed on page 85 and, for freeze-avoiding species, on page 103.

Ice nucleating agents

The insect body can be regarded as a series of fluid-filled compartments. The temperature at which the contents of these fluid-filled compartments will freeze varies, as does the amount of injury that will result. Most of these compartments, such as the digestive system and individual cells, contain *nucleators* – components with the ability to induce freezing at temperatures only a few degrees below the melting point of the contents. Nucleators in the cells and intestine are common to both freeze tolerant and intolerant insects but have to be evacuated or 'masked' in winter in the latter group to maximise supercooling. Most freeze tolerant insects actively synthesise ice nucleators in the haemolymph during autumn and early winter and the efficiency of these agents compared to others in the body is central to the success of the tolerance strategy. The synthesised ice nucleators provide a surface upon which an embryo crystal can grow to a critical size at a temperature higher than that required in the absence of the nucleating agent (Duman *et al.* 1984).

The structure, function and mechanisms of ice nucleating agents have been reviewed by Zachariassen (1982, 1985). The physiological function of these nucleators in freeze tolerant insects is probably to prevent freezing in closed compartments such as the cells and intestine, where ice formation is likely to cause severe injury or death. It seems likely that some of these compartments, such as cells, contain nucleators throughout the year, whereas in the digestive system there may be an elimination of potential nucleating agents in those species that overwinter in a starved condition. If freezing occurs in a cell there will be an increased solute concentration around the ice crystal and water will move by osmotic influx from the extracellular spaces into the cell, resulting in progressive swelling of the cell and eventually a rupture in the cell membrane. By contrast, if freezing is first initiated in extracellular areas, there will be an osmotic gradient across the cell membrane and a gradual movement of water from the cells that will undergo a 'shrinking' process as the extracellular ice masses grow in size. If a vapour pressure equilibrium is maintained between the intra- and extracellular fluid and extracellular ice, there is little risk of spontaneous freezing in any cell, or any other closed compartment functioning in the same way. The lipid plasma membrane of cells and the intestine prevents the extension of extracellular ice into the compartments. This theory owes much to research on plant cold hardiness described in Levitt (1980).

A primary role of ice nucleating agents in freeze tolerant insects is therefore to ensure that extracellular freezing is initiated before other nucleators become operative in areas of the body where freezing would be lethal. Zachariassen and Hammel (1976) demonstrated the seasonal occurrence and activity of haemolymph ice nucleators in the freeze tolerant beetle *Eleodes blanchardi* (Coleoptera: Tenebrionidae).

Table 4.1. *Species of freeze-tolerant insects with nucleating agents in the haemolymph (after Zachariassen, 1982)*

		Supercooling points (°C)	
Species	Location	Intact insect	Haemolymph +0.9% NaCl
Coleoptera			
Phosphuga atrata	Norway	−5.5	−5.6
Pytho depressus	Norway	−5.0 ± 0.2	−8.9 ± 1.2
Eleodes blanchardi	S. California, USA	−6.3 ± 0.6	−5.7 ± 0.3
Eleodes humeralis	Washington, USA	−6.1	−6.1 ± 0.3
Coelocnemis californicus	S. California, USA	−6.8 ± 0.4	−5.9 ± 0.2
Coelocnemis magna	S. California, USA	−6.4	−6.3 ± 0.6
Melasoma collaris	Norway	−4.7	−6.3 ± 0.5
Phyllodecta laticollis	Norway	−5.0	−7.3
Seneciobius kenyanus	Mount Kenya	−5.2 ± 1.2	−6.3 ± 0.0
Parasystates elongatus	Mount Kenya	−5.2 ± 1.3	−7.3 ±
*Dendroides canadensis**	Indiana, USA	−7.0	−
Hymenoptera			
Vespula maculata	Indiana, USA	−4.6 ± 0.8	−4.1 ± 0.3†
Diptera			
*Eurosta solidaginis**	Alberta, Canada	−10.1 ± 1.1	−9.8 ± 1.4

* *larvae*; † diluted 1:100 in distilled water

Apparently the nucleators were only present in freeze tolerant winter specimens with higher supercooling points than summer-collected beetles, which were also freezing intolerant. Subsequently, haemolymph ice nucleators have been found in a number of freeze tolerant beetles as well as in Hymenoptera (Table 4.1) and Zachariassen (1980) has suggested that the presence of such nucleating agents is a characteristic feature of adult freeze tolerant beetles. Although the great majority of freeze tolerant insects have elevated supercooling points, in some species the nucleators appear to be active in sites other than the haemolymph. For instance, the whole intact larva of *Eurosta solidaginis* (Diptera: Tephritidae) freezes at −9 °C, but the larval haemolymph does not freeze until −17 °C (Bale *et al.* 1989a); Sømme (1978) recorded similar supercooling points in the larva and haemolymph of *E. solidaginis*. It is possible that some species, such as *E. solidaginis*, may have more than one site of nucleating, both (or all) of which promote safe extracellular freezing at around −9 °C. It is interesting to note that, while freezing in the intestinal tract is regarded as lethal to freezing intolerant insects (most of which evacuate the digestive system in autumn), a recent report (Shimada 1988) has suggested that the gut, and not the haemolymph, is the site of ice nucleation and safe freezing in the freeze tolerant prepupa of *Trichiocampus populi* (Hymenoptera: Tenthredinidae).

The data presented in Table 4.1 indicate the limited supercooling ability of freeze tolerant insects, most of which have winter supercooling points between -5 and $-10\,°C$, and the similarity between the supercooling points of the haemolymph and the whole intact insect. In this respect, an alternative but compatible theory on the function of ice nucleators in freeze tolerant insects has been put forward by Duman and Horwath (1983) who argued that extracellular ice nucleators act to limit the amount of supercooling in the insect. Supercooling in a freezing intolerant insect is, by necessity, extensive but ice crystal growth at the point of spontaneous nucleation is extremely rapid, such that the diffusion of cell water to extracellular ice cannot occur with sufficient speed to prevent intracellular freezing. It seems reasonable to conclude that ice nucleating agents in freeze tolerant insects have at least a dual function – the initiation of extracellular freezing and the limitation of supercooling.

The action of potent haemolymph ice nucleators provides the added advantage that less efficient nucleators in the cells and intestine do not have to be masked or removed in winter and intermittent freeze–thaw cycles can be tolerated provided that nucleators retain their activity throughout winter (Zachariassen 1985). In some environments, insects may experience subzero temperatures during the summer season at a time when feeding, reproduction and development must occur. For instance, adult chrysomelid beetles, *Phyllodecta laticollis* (Coleoptera: Chrysomelidae) (Van der Laak 1982), *Seneciobius kenyanus* (Coleoptera: Curculionidae) and *Parasystates elongatus* (Coleoptera: Curculionidae) (Sømme and Zachariassen 1981) live in habitats where there can be a diurnal change in temperature in summer from $10\,°C$ in the daytime when the insects feed to $-10\,°C$ overnight. In these species, the risk of 'night-time' freezing induced by food particle nucleators ingested in daytime feeding is overcome by the action of more effective haemolymph ice nucleators, which result in extremely high summertime supercooling points.

The structure of haemolymph ice nucleators has been investigated in *Eleodes blanchardi* by Zachariassen and Hammel (1976) and in queens of the bald-faced hornet, *Vespula maculata* (Hymenoptera: Vespidae), by Duman and Patterson (1978). Heating the haemolymph to $+100\,°C$ inactivated the nucleators in both species and $+80\,°C$ was sufficient to inactivate nucleating agents in *Rhagium inquisitor* (Coleoptera: Ceramby-cidae) (Baust and Zachariassen 1983); additionally protease enzymes inactivated the *Vespula* nucleators. Collectively, this information was taken to indicate that the nucleators were proteinaceous. More recently, Duman *et al.* (1984) have purified an ice nucleating protein from *Vespula*. Using sodium dodecyl sulphate–polyacrylamide gel electrophoresis (SDS–PAGE), the protein was found to have a molecular mass of 74 000 Da. No sugars or amino sugars were detected and it was assumed that the nucleator was a pure protein. The protein consisted of 18 different amino acids and the most abundant – glutamate (or glutamine) (20 per cent), serine (12 per cent) and threonine (11 per cent) – all have hydrophilic properties. Duman *et al.* (1984) suggested that the general hydrophilic nature of the protein may confer considerable ability to order water and increase the likelihood of nucleation.

It is evident that most freeze tolerant insects synthesise or unmask haemolymph nucleators in winter to induce extracellular freezing and minimise supercooling. However, it is a feature of research on insect cold-hardiness, and indeed many areas of biological science that, as investigations become more detailed and cover a wider range of subject species, examples come to light that do not fit the general rule. Ring (1982) described a comparative study in Canada on the alpine populations of the overwintering freeze tolerant larvae of *Pytho deplanatus* (Coleoptera: Pythidae) and a congeneric species, *P. americanus*. The larvae of both species overwinter under the bark of fallen spruce trees, which in winter are covered by an insulating layer of snow. Under a laboratory acclima-tion regime in which the temperature was lowered from an initial $+8\,^{\circ}\text{C}$ stepwise to $-15\,^{\circ}\text{C}$ and back to $-5\,^{\circ}\text{C}$ over 36 weeks, the supercooling points of *P. americanus* were consistently within the range of -3.5 to $-6.6\,^{\circ}\text{C}$, values typical of freeze tolerant insects with effective haemo-lymph ice nucleators. By contrast, by week 25 (when the larvae were held at $-15\,^{\circ}\text{C}$) the mean supercooling point of *P. deplanatus* had decreased to $-54\,^{\circ}\text{C}$, the lowest recorded supercooling point for a freeze tolerant insect. The physiological and biochemical basis for this phenomenon involving an increased solute (glycerol) concentration and a much reduced water content will be discussed in the next section. It seems that freeze tolerant insects with such low supercooling points are unable to synthesise nucleators but are able to eliminate or mask other potentially lethal nucleating agents in the gut and other fluid-filled compartments. Ring (1982) suggested that the strategy had an adaptive advantage in that larvae would normally overwinter in a supercooled state, but in the event

Table 4.2. *Species of freeze-tolerant insects with low supercooling points*
(*after Ring 1982*)

Insect	Stage	Supercooling point (°C)*	Cryoprotectant (%)
Coleoptera			
Pytho deplanatus	Larva	−54	Glycerol (13%)
Mordellistena unicolor	Larva	−26	−
Lepidoptera			
Polygonia sp.	Adult	−25	Glycerol (9%)
Martyrhilda ciniflonella	Adult	−23	Glycerol (8%)
Diptera			
Mycetophilia sp.	Adult	−33	Glycerol (14%)
Exechia sp.	Adult	−33	Glycerol (14%)
Hymenoptera			
Bracon cephi	Larva	−47	Glycerol (29%)
Eurytoma gigantea	Larva	−49	Glycerol (23%)
Eurytoma gigantea	Larva	−27	−
Eurytoma obtusiventris	Pupa	−29	−
Rogus sp.	Adult	−30	Glycerol −

* lowest recorded value in winter or during laboratory acclimation.

of freezing, the antifreeze function of the high glycerol content and blood sugars would assume a cryoprotective role. This view has been supported by Zachariassen *et al.* (1979), who have shown that injuries develop in freezing intolerant insects when about 50 per cent of the body water is frozen, whereas the glycerol concentrations in freeze tolerant insects such as *P. deplanatus*, with supercooling points below −50 °C, would limit ice formation to less than half the body water. Additionally, Baust (1973) has demonstrated that the configuration of ice crystals in concentrated glycerol solutions would make them less able to penetrate cell membranes. A number of other freeze tolerant insects have been found with low supercooling points and these are summarised in Table 4.2 (Ring 1982).

Cryoprotectants and cryoprotection in freeze tolerant insects

Most insects that overwinter in a freeze tolerant state accumulate polyols and carbohydrates of which glycerol is overwhelmingly the most common and abundant substance (Sømme 1964; Baust 1973; Zachariassen 1985). Other polyols recorded in freeze tolerant species include sorbitol, threitol, erythritol and the sugars fructose, sucrose and trehalose.

Apart from the action of ice nucleators to limit winter supercooling, most freeze tolerant insects share a number of other common features. Their cold-hardiness increases greatly in winter, reflected by the acquisition of the tolerant state and lowering of the lethal temperature, and is accompanied by the accumulation of cryoprotectants that are absent or found at very low levels in summer. Polyols and other cryoprotectants are more effective in promoting cold-hardiness in freeze tolerant insects than those that depend upon supercooling for survival, allowing freeze tolerant species to overwinter in extreme environments where freezing intolerant species would rapidly perish. The section below gives examples of 'case-history' studies on some freeze tolerant species and this information provides a basis for discussing the various interactions between assessments of cold-hardiness and cryoprotectant concentration and theories on the mechanisms of cryoprotection.

The carabid beetle, *Pterostichus brevicornis* (Coleoptera: Carabidae), was studied in Alaska, where it overwinters as adults and larvae. Adult beetles seek out decaying tree stumps and felled timber as overwintering sites and, although the habitat is at times partly covered with snow, the beetles may experience temperatures as low as −60 °C, with prolonged exposure over several weeks to −40 °C (Miller 1969; Baust and Miller 1970). Measurements made on seasonally collected beetles and summarised in Figure 4.2 (Baust and Miller 1970) showed that:

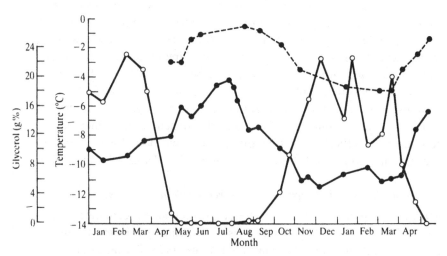

Fig. 4.2. Seasonal variation in haemolymph glycerol content, supercooling points and haemolymph freezing points in adult individuals of the carabid beetle, *Pterostichus brevicornis*. ○—○ glycerol, ●—● supercooling points, ●---● haemolymph freezing points (after Baust and Miller 1970).

1. Glycerol content increased from less than 1 per cent in spring and summer to more than 22 per cent in winter. Glycerol synthesis was temperature dependent and increased after the first frost. The haemolymph contained trehalose but no other polyols.
2. Supercooling points decreased from $-6\,^{\circ}C$ in summer (when freezing was fatal) to $-11\,^{\circ}C$ in winter, when the beetles became freeze tolerant. (In subsequent experiments, Baust and Morrisey (1977) found that 10 per cent ($\sim 1\,M$) was a critical threshold below which *Pterostichus brevicornis* was freezing intolerant.)
3. Haemolymph freezing points also varied seasonally from $-0.6\,^{\circ}C$ in summer to $-5\,^{\circ}C$ in late winter.

Although the depression of the supercooling points and haemolymph freezing points correlates with the increasing glycerol content (Fig. 4.2), the increase in supercooling from summer to winter of $5\,^{\circ}C$, coupled with a depression of the haemolymph freezing point to only $-5\,^{\circ}C$, suggests that supercooling was effectively limited by the action of winter active ice nucleating agents, as recognised by Miller (1982). In the freeze tolerant state, most overwintering *P. brevicornis* can survive to $-70\,^{\circ}C$ and some individuals to $-87\,^{\circ}C$ (Miller 1969). Ice nucleators are active in the majority of freeze tolerant insects only in winter, and the species are freezing intolerant at other times of the year. This change does not generally present any risk because potentially lethal subzero temperatures are not encountered during the late spring–summer–early autumn period when the insects are unable to tolerate freezing.

In the section on ice nucleating agents, mention was made of freeze tolerant species that may experience subzero temperatures throughout the year and some of these species are now known to retain their freeze tolerant condition all year round. One such species, the chrysomelid beetle, *Phyllodecta laticollis*, overwinters in Norway as adults in exposed sites on or beneath the loose bark of trees, where the winter temperature may fall to -30 to $-40\,^{\circ}C$; *P. laticollis* has been the subject of a comprehensive study by van der Laak (1982).

The supercooling points of the beetles varied little, from $-5\,^{\circ}C$ in summer to $-7\,^{\circ}C$ in winter (Fig. 4.3) suggesting that some types of nucleators were active throughout the year. The supercooling points of 0.9 per cent NaCl solution, to which haemolymph samples were added from beetles collected in autumn, winter, spring and summer were also highly consistent (range -7.1 to $-7.7\,^{\circ}C$), confirming that haemolymph nucleators were present at all seasons. The low temperature tolerance

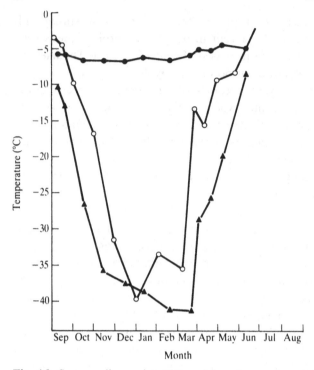

Fig. 4.3. Supercooling point (●) and low temperature tolerance limit (▲) of *Phyllodecta laticollis* and minimum temperature (○) at the collection site during the year (after van der Laak 1982).

varied from −9 °C in summer to around −42 °C in winter, such that the lower lethal limit was continuously below the lowest environmental temperatures, apart from a brief period in midwinter.

The increased cold-hardiness of the beetles from summer to winter correlated well with an accumulation of glycerol from zero in midsummer to around 1500 mmol/kg in winter with a parallel change in the haemolymph osmolality from 500 mosM to about 2500 mosM (Fig. 4.4). The osmotic contribution of glycerol accounted for almost the entire increase in the haemolymph osmolality. Although the changes in cold-hardiness in the phases of autumn acclimation and spring/summer deacclimation followed a pattern of increasing and decreasing glycerol concentration and haemolymph osmolality, a marked shift in the relationship was noted between autumn and spring. For instance, in October, at a haemolymph osmolality of about 1800 mosM the lower lethal limit was around −30 °C, whereas in spring at the same osmolality the low temperature tolerance was around −40 °C.

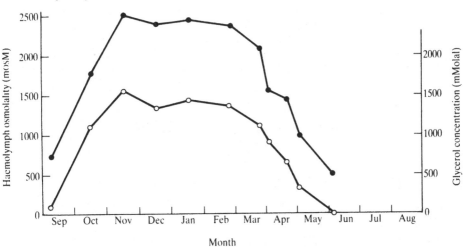

Fig. 4.4. Haemolymph osmolality (●) and glycerol concentration (○) in *Phyllodecta laticollis* during the year (after van der Laak 1982).

Phyllodecta laticollis thus shows two advantageous adaptations; first, the beetles are freeze tolerant throughout the year and second, they show an increased freezing tolerance in spring compared to autumn. The freezing tolerance in summer is of great importance, because at that time of year the beetles feed on *Salix* leaves, staying on the leaves overnight, when the temperature may fall to −5 °C. Without the haemolymph nucleators, freezing would occur first in the gut and would almost certainly be fatal. The relative increase in cold-hardiness in spring is also beneficial in a climate where periods of mild weather are likely to be followed by sudden changes to severe frosts. Periods of warm weather and daytime solar warming of the habitat may reduce the glycerol content so that the ability to tolerate lower temperatures in spring than autumn at the same glycerol concentration is clearly an aid to survival.

In contrast to *Pterostichus brevicornis* with its winter active ice nucleators, and *P. laticollis*, with continual protection from haemolymph nucleative agents, other species, such as the freeze tolerant larvae of *Pytho deplanatus*, lack haemolymph ice nucleators in winter and supercool extensively, to considerably below the supercooling points of many freezing intolerant species (Ring and Tesar 1981; Ring 1982). In a laboratory cold acclimation regime, the water content of *P. deplanatus* decreased from 68.8 per cent in non-acclimated larvae to 30 per cent when larval cold-hardiness was at a maximum (based on lower lethal

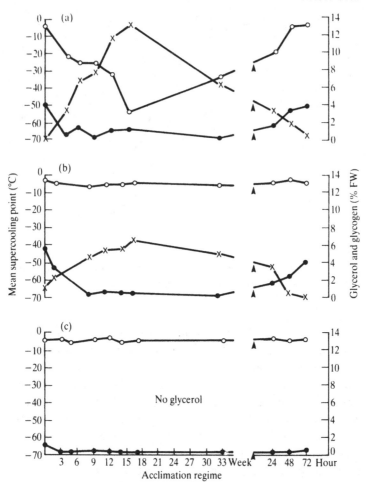

Fig. 4.5. Supercooling points (○), glycerol (×) and glycogen (●) concentration during low temperature acclimation and deacclimation in the larvae of (a) *Pytho deplanatus*, (b) *Pytho americanus* and (c) *Xylophagus* sp. ▲ indicates the beginning of rewarming period at 22 °C (after Ring 1982).

temperatures) after a stepwise reduction in temperature over many weeks to −15 °C (Ring 1982). The lowest supercooling point of −54 °C was observed after 14 weeks exposure to subzero temperatures, at which time glycerol and sugars had reached their maximum concentrations of 13.2 and 5.5 per cent fresh weights, respectively, and glycogen content was also reduced at 1.2 per cent (compared to 4.1 per cent fresh weight in unacclimated larvae) (Fig. 4.5a). In this condition larvae survived to −55 °C with no visible effects of freezing damage.

In comparative studies on *P. americanus*, which also overwinters as freeze tolerant larvae under fallen timber, the water contents of non-acclimated larvae and cold-hardy larvae exposed to a similar cold acclimation regime as *P. deplanatus*, were 65 and 66 per cent, respectively. Additionally, in the change from the non-acclimated state to the phase of maximum cold-hardiness, the supercooling point decreased from $-3.5\,°C$ to only $-5\,°C$ (a clear indication of effective ice nucleators), glycerol content (fresh weight) increased from 1.1 to 6.6 per cent and glycogen decreased from 5.6 to 0.41 per cent (Fig. 4.5b). In this cold-hardy phase larvae could survive in the frozen state to below $-40\,°C$.

Although the glycerol content expressed as a percentage of the fresh weight was approximately double in *P. deplanatus* compared to *P. americanus* during the period of maximum cold-hardiness, the water content of *P. deplanatus* was only half that of *P. americanus*, emphasising the difficulty highlighted by Zachariassen (1985) in comparing cryo-protectant concentrations when expressed as units of solute per unit of fresh body weight, in species where the relative water content may vary substantially.

P. deplanatus represents a small group of freeze tolerant larvae with low supercooling points, based on the absence in winter of haemolymph nucleators, and a marked increase in the concentration of polyols and sugars. In the opinion of Ring (1982) the supercooling ability of *P. deplanatus* was so extensive that overwintering larvae were unlikely to freeze, but in the event of nucleation the 'antifreeze' polyols and sugars would then assume a cryoprotective role.

The studies on alpine populations of *P. deplanatus* and *P. americanus* (the latter also having an arctic distribution) included an unrelated *Xylophagus* species (Diptera: Xylophagidae), which shares an arctic distribution with *P. americanus* and also overwinters as freeze tolerant larvae in the same niche. In common with *P. americanus* the water content remained consistent throughout the acclimating regime at 60–65 per cent, the supercooling point varied little (from -4.8 to -5 to $-6\,°C$), suggesting that nucleators were active, and the glycogen content decreased. While the lower lethal limit of *Xylophagus* is similar to both *Pytho* species, the freeze tolerant survival of the former is achieved without any change in glycerol content (which was recorded as an unquantifiable trace in all larvae) or significant amounts of other polyols or sugars (Fig. 4.5c) (Ring 1982). One of the few adaptations shared by all three species was the rapid deacclimation response in which warming for three days at $22\,°C$

raised the lower lethal limit by 30–40 °C, with the virtual loss of glycerol in *P. deplanatus* and *P. americanus.*

In view of the tendency of insect cryobiologists to generalise on the principles of insect cold-hardiness in the search for some common theory of cold tolerance, it is salutary to reflect on the overwintering strategies of the two *Pytho* species and *Xylophagus.* All three species are freeze tolerant, survive to −40 °C and below, and overwinter in the same type of habitat. *P. deplanatus* reduces its water content by half but *P. americanus* and *Xylophagus* do not. *P. deplanatus* and *P. americanus* accumulate glycerol, but *Xylophagus* does not. *P. americanus* and *Xylophagus* appear to contain winter active ice nucleators, but *P. deplanatus* does not!

Clearly, species such as *Xylophagus*, which show no obvious cryo-protectants but relatively low lethal temperatures, are rare. But this diversity in the biochemistry of overwintering insects, the presence or absence of nucleators, the variation in number and concentration of polyols and sugars, suggests that it is unlikely that there is a single 'mechanism of cryoprotection' in freeze tolerant insects. This view was recognised as early as 1973 by Baust, who listed the range of theories on freezing damage (mechanical damage at site of freezing, electrolyte imbalance, critical cell volumes and recrystallisation) and then suggested possible roles for polyols such as glycerol in counteracting *all* of the deleterious effects described (Baust 1973). For instance, in addition to their solute (colligative) *antifreeze* effect, which lowers haemolymph and whole body freezing (supercooling) points and is therefore the primary strategy for survival in freezing intolerant insects, the cryoprotective functions of these compounds include:

1. An increase in the water binding capacity (attributable to their polyhydroxy nature) and reduction in the rate of ice spread and the total ice content following freezing; the lower ice content would reduce the likelihood of intracellular freezing.
2. Stabilisation of protein structure and buffering of electrolytes.
3. Reduction in transmembrane water fluxes and maintenance of cell volumes above critical minima.
4. Decreases in the probability of recrystallisation.
5. Maintenance of membrane fluidity and decreased sensitivity to changes in ambient freezing rates (Baust 1973, 1982).

In the light of this multifunctional concept of cryoprotectants, it must be expected that observations made of different species will vary (based

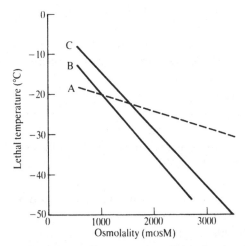

Fig. 4.6. Lethal temperatures of freezing intolerant insects (Line A) and freezing tolerant *Phyllodecta laticollis* (Line B, Van der Laak 1982) and *Pytho depressus* (Line C, Zachariassen 1979) (after Zachariassen 1985).

mainly on the different biochemical profiles of the species) and provide evidence in support of particular 'theories of cryoprotection'. But these differences should not be regarded as necessarily contradictory or mutually exclusive; they may simply reflect the obvious diversity in the form and function of insects.

The common occurrence of polyols and sugars in freezing tolerant and intolerant species, often at similar concentrations, allows their cold-hardiness effect to be compared between the two strategies on the basis of lower lethal temperatures. As shown in Figure 4.6 at concentrations above 1000 mosM/kg (*Phyllodecta laticollis*; van der Laak 1982) and approximately 1500 mosM/kg (*Pytho depressus*; Zachariassen 1977, 1979), freeze tolerant insects of this type are increasingly more cold-hardy than freezing intolerant species with the same haemolymph osmolality – based on the regression lines in Figure 4.6, at 2500 mosM/kg the lower lethal temperature of *P. laticollis* is about −43 °C and *P. depressus* about −36 °C, whereas the supercooling point of a range of freezing intolerant species at the same concentration is about −25 °C. These results were taken to indicate that freeze tolerant insects were able to utilise their carbohydrate resources more effectively than freezing intolerant species and that freeze tolerance was likely to be the more successful strategy in areas with extremely low temperatures (Zachariassen 1980, 1985). These ideas are borne out by the studies of Miller (1982) and Ring (1982) who described many freeze tolerant species in interior Alaska and Canada respectively, some of which could survive below −80 °C.

The discovery of antifreeze proteins in freeze tolerant insects has led to speculation on their role in these species because their antifreeze properties might be expected to reduce the efficiency of ice nucleating proteins, which are crucial to the success of overwintering in the freeze tolerant state. It has been suggested that antifreeze proteins may be important in autumn prior to the development of freeze tolerance, and again in spring after freeze tolerance has been lost; antifreeze proteins would provide sufficient antifreeze protection against early and late frosts (Duman 1979, 1980; Duman *et al.* 1982).

An additional, and perhaps more important function, has emerged from studies on antifreeze proteins in arctic fish. At very low concentrations, antifreeze glycoproteins from the blood plasma of the fish are highly effective inhibitors of ice recrystallisation (Knight *et al.* 1984); similar activity in insect antifreeze proteins would aid survival in freeze tolerant insects that freeze at elevated temperatures but have to survive at −30 to −60 °C in nature, and may be at risk of damage from recrystallisation during periods of warming to higher subzero temperatures.

Vitrification

Once freezing has been initiated in a freeze tolerant insect at −5 to −10 °C, water moves from cells to extracellular areas under an osmotic gradient, resulting in a cryoconcentration of the cell contents. This process will continue as long as the temperature is lowered and the ice masses continue to grow. In its frozen state the insect will contain extracellular ice and unfrozen extracellular and intracellular fluids, which will move toward osmotic equilibrium. A recent study (Wasylyk *et al.* 1988) has shown that artificial haemolymph containing the concentrations of polyols and sugars (determined by HPLC) found in freeze tolerant larvae of *E. solidaginis* is capable of partial vitrification (glassy state formation) following extracellular ice growth to approximately −25 °C.

Numerous substances have the ability, when cooled and with increased viscosity, to form an amorphous solid or glass. The vitrified liquid assumes a solid state characterised by a highly viscous non-crystalline solid that lacks transitional motion at the atomic level; in simple terms, molecular diffusion stops on a biological time scale.

Prior to the study by Wasylyk *et al.*, glass formation in aqueous solutions had been reported only in multimolar solutions. However, *Eurosta* produces a submolar, multicomponent cryoprotectant system (glycerol, sorbitol, fructose and trehalose) that becomes glassy at a relatively high subfreezing temperature (-25 to $-35\,°C$). The concept of intracellular vitrification in insects is attractive, explaining some of the paradoxes not often addressed in the literature and conferring many biological advantages to aid survival. For instance, ice growth is terminated upon vitrification, preventing further solute concentration and cell shrinkage and, because of the multicomponent system, toxic levels associated with single cryoprotectants are not reached. Importantly the probability of intracellular freezing is reduced to zero. Additionally, coincidental diffusion of ions is suppressed and the solubilisation of intracellular adenylates, such as ATP in the presence of glycerol, is eliminated. Vitrification would also protect macromolecular fine structures through the maintenance of adequate water hydration and, because of the proximity of the vitrification temperature to the melt temperature, devitrification (secondary crystallisation) is unlikely to occur upon warming. Furthermore, insect cells would not experience the osmotic stresses predicted for non-glassy solutions at the same concentration.

It seems that insects have evolved a survival strategy similar to that routinely employed for the cryopreservation of mammalian tissues at liquid nitrogen temperatures ($-196\,°C$) but achievable at a much higher temperature and sustainable over the winter period (J. G. Baust, personal communication).

Freeze avoidance by supercooling

Freezing intolerant insects predominate in cool and cold climates; freeze tolerant species are more abundant in regions where winter frosts can be extreme. The majority of cold-hardy insects are freezing intolerant and rely on the process of supercooling to avoid lethal freezing.

The melting point of pure water is $0\,°C$, but when a sample is cooled it will not normally freeze at its melting point but will remain unfrozen until cooled to much lower temperatures. Liquids that remain unfrozen at temperatures below their melting point are described as 'supercooled' and the temperature at which spontaneous freezing occurs is the 'crystallisation temperature', which is commonly described as the 'supercooling point'. For instance, small volumes of highly purified water will supercool to around $-40\,°C$ (MacKenzie 1977). In simple terms the insect body can be viewed as a liquid container in which the conditions for supercooling are exceptionally favourable (Sømme 1982). In most freezing intolerant insects this favourable state is brought about by behavioural and physiological adaptations. First, feeding usually ceases in the autumn and the digestive system is cleared of any food particles that might act as nucleation sites. At the same time, other nucleators that cannot be eliminated from the body must be inactivated or 'masked'. In some insects this action alone may be sufficient to allow supercooling to below $-20\,°C$. Second, in the great majority of freezing intolerant insects there is a seasonal (winter) increase of antifreeze agents, notably polyols and proteins. When a solute such as glycerol is added to water both the melting point and supercooling point are depressed, but not by equal amounts. Various studies (Sømme 1967; Block and Young 1979; Zachariassen 1980) have shown that, at a given solute concentration, the depression of the supercooling point is more than twice the corresponding depression of the melting point. The accumulation of polyols in winter therefore increases the supercooling capacity of the nucleator-free liquid compartments in the insect and lowers the temperature at which lethal freezing occurs. The importance of polyols in freezing intolerant insects, the role of antifreeze proteins and the influence of environmental factors on supercooling are considered in the following sections.

Measurement of supercooling points

The limit of supercooling in terrestrial arthropods is conveniently measured as the supercooling point (Sømme 1982). As freezing of some part of the body is initiated at the supercooling point (and proceeds to other regions if the temperature is held constant or is further decreased) and such freezing is, by definition, lethal to freezing intolerant insects, the super-cooling point has been widely used as an indicator of the lower lethal temperature of insects that try to overwinter by a freeze avoidance strategy. The supercooling point is the temperature at which spontaneous freezing occurs during gradual cooling and is detected as the latent heat emitted during the transition of water to ice.

Equipment for measuring supercooling should cool the specimen at a constant and controllable rate and detect the latent heat emitted when ice forms in the animal. A standard cooling rate of 1 °C/min was proposed by Salt (1966a) and virtually all supercooling studies have been carried out in the range of 0.5–2 °C/min; adoption of a standard rate was recommended to allow comparisons between species and investigators. The effect of cooling rate on the measured supercooling point and the fate of an insect on freezing is a matter of considerable debate at present (see p. 139). If the cooling rate does affect the level of supercooling and if the cooling and warming rates affect the ratio of deaths to survivors, implying that the optimal rate for survival may differ between species, then clearly a system for measuring supercooling must provide a wide range of cooling and warming profiles with constant and variable rates between the starting temperature and the supercooling point.

The most commonly used cooling method is the 'alcohol bath', in which specimens are attached to thermocouples and placed in small glass tubes suspended in alcohol (Block and Young 1979; Ring and Tesar 1980). Modern systems of this type incorporate a temperature programmer to achieve a defined rate of cooling. Some investigators have used 'freezers' (Cloudsley-Thompson 1973) but these are subject to considerable error. A freezer operates continuously at subzero temperatures and usually cycles around a mean temperature. When an insect is first placed in a freezer the rate of heat loss can be rapid, but becomes increasingly slower as the specimen approaches equilibrium with the freezer temperature; the rate of cooling is therefore uncontrolled and the minimum temperature of some freezers (e.g. $-20\,°C$) is insufficient to obtain supercooling points for cold-hardy insects.

Another system of cooling utilises fluid-cooled thermoelectric modules (based on the Peltier effect) and these offer programmable rapid, slow or 'stepped' cooling rates when linked to an automatic controller (Bale *et al.* 1984). Thermocouples are the preferred temperature sensors in most studies because they are available in sufficiently small size to detect the heat emission of small insects on freezing. The thermocouples (copper–constanton) are usually connected via ice point references to multichannel chart recorders or to recorders with an internal cold junction compensation. A system has recently been developed to interface thermocouples via a thermocouple converter to a microcomputer (Bale *et al.* 1984). Under this arrangement, individual thermocouples can be 'matched' to a high level of accuracy ($\pm 0.2\,°C$) by incorporating polynomial descriptor equations for each sensor in the computer program; an added advantage is the minimal influence of changes in ambient on the accuracy of the thermocouples ($1\,°C$ for $50\,°C$ shift in ambient).

For some applications, such as research on plant parasitic nematodes with a fresh weight of only $0.1\,\mu g$, the heat emission on freezing is insufficient for detection by a thermocouple. For such very small invertebrates thermistors may be more appropriate and multichannel thermistor units, which are compatible with any method of cooling, are now available (Knight *et al.* 1986a).

Supercooling in freezing intolerant insects

Sømme (1982) provides a detailed account of the occurrence of super-cooling in eggs, larvae (nymphs), pupae and adult insects. To survive as a species a proportion of the overwintering population, whatever the stage or instar, must be able to survive the minimum winter temperatures of the selected overwintering site. While it may be unreliable, it is undoubt-edly convenient to assess the survival potential of a species by comparing its measured supercooling point (as an indicator of the lower lethal limit) to minimum winter temperatures. Studies on supercooling are described extensively in this chapter and three major limitations on their usage as an indicator of the low temperature death point in nature (which will be discussed later in this chapter and elsewhere in this book) should therefore be mentioned at this stage: Firstly, supercooling points are measured at a uniform rate of cooling (usually 1 °C/min) and reflect supercooling ability at a fixed moment in time. In nature rates of cooling and warming are variable and almost certainly slower than those used in the laboratory and in winter the insect may be exposed to subzero temperatures continuously for long periods of time. Secondly, winter temperatures of different climates are often based on some standard measurement, such as a meteorological screen, whereas insects may select overwintering sites where the microclimate temperature is frequently above (soil) or below (field surface) the standard measurement. Thirdly, the use of supercooling points as a direct measure of the lower limit assumes that: (i) freezing is the cause of death; and (ii) the insect remains alive and is capable of recovery as long as the supercooling point is not exceeded (Bale 1987). In fact some insects, such as aphids, are killed or fatally injured during cooling at 1 °C/min at temperatures 10–15 °C above their supercooling point (Knight *et al.* 1986b; Bale *et al.* 1987). Results of this type indicate that assessments of pre-freeze mortality should be included as a matter of routine in supercooling studies on apparently freezing intolerant insects.

With these limitations in mind it is apparent that, in a comparison of overwintering stages of a large number of species, eggs are often more cold-hardy (with lower supercooling points) than larvae, pupae or adults (Sømme 1982). For instance, eggs of aphids and psyllids supercool to −40 °C (Sømme 1964; Skanland and Sømme 1981; James and Luff 1982) and supercooling points of −50 °C have been recorded in some lepi-dopteran eggs (Bakke 1969). Eggs are likely to be free of nucleators and are therefore extremely efficient vessels for supercooling. High super-cooling ability has been found in the eggs of species that would not

normally experience adverse conditions of low temperature, such as
− 30 °C in *Locusta migratoria* (Orthoptera: Acrididae) and this suggests
that extensive supercooling may be a natural property of insect eggs.
Many species deposit overwintering eggs in highly exposed sites on bushes
and trees and low supercooling points are therefore a prerequisite for
winter survival.

Most overwintering larvae supercool to − 20 or − 30 °C (Sømme 1982)
and this level of cold-hardiness in combination with the protection
afforded by overwintering sites in the soil and under bark or snow, may
be sufficient to ensure survival. A number of species with overwintering
larvae have very low supercooling points such as − 53 °C in the bark
beetle, *Scolytus multistriatus* (Coleoptera: Scolytidae) (Lozina-Losinkii
1974) and − 66 °C in individuals of a *Rhabdophaga* (Diptera: Cecidomyiidae) species (Ring and Tesar 1980), which forms galls on willow.

Less information is available on overwintering pupae than on other
stages, but the majority of species studied, most of which are Lepidoptera,
have supercooling points between − 20 and − 30 °C. Pupal cocoons are
often constructed in the soil, which acts as an extremely effective buffer
against severe frosts at the surface. In England, for instance, overwintering
pupae of cabbage root fly *Delia radicum* (Diptera: Anthomyiidae) with a
mean winter supercooling point of − 23 °C, overwinter at about 5 cm soil
depth and may not experience subzero temperatures at any time throughout an entire winter (Block *et al.* 1987), although there may be more than
100 frosts a year at the soil surface (Bale *et al.* 1988).

The supercooling points of overwintering adults vary considerably
between species and there is no consistent pattern, even within an order.
Many Coleoptera in West Germany have winter supercooling points
about − 10 °C (Topp 1978) and values between − 8 and − 18 °C were
found in five beetle species from Mount Kenya (Sømme and Zachariassen
1981). A number of adult beetles have supercooling points between − 20
and − 30 °C, such as the beech weevil, *Rhynchaenus fagi* (Coleoptera:
Curculionidae) (Bale 1980) and only a few species of overwintering adults
are able to supercool to below − 30 °C, e.g. *Rhagium inquisitor* at − 35 °C,
although Zachariassen and Husby (1982) recorded a mean of − 27 °C for
the same species. Sømme (1982) noted that many adult insects overwinter
in soil, leaf litter, tree stumps or rotten wood and Danks (1978) emphasised the need to distinguish between the microhabitat temperatures
experienced by the insects and standard air temperature measurements,
which are often used to categorise countries or regions. In Britain, a
consistent winter supercooling ability of only − 10 °C would be an

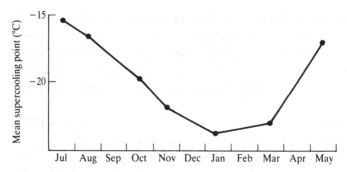

Fig. 4.7. Seasonal changes in the supercooling of the beech weevil, *Rhynchaenus fagi.*

adequate protection against freezing for any insect overwintering at 10 cm soil depth.

Seasonal changes in supercooling

The most striking feature of freezing intolerant insects is the seasonal increase in supercooling from summer through autumn to winter. The process is described as acclimatisation (natural field changes), acclimation (laboratory-induced) or simply cold-hardening. The mean supercooling point of unfed adult beech leaf mining weevils, *Rhynchaenus fagi*, dissected from their pupal cocoons in late June was − 15.4 °C (Fig. 4.7; Bale 1980); by October this had decreased to − 19.3 °C and then to − 23.1 °C in early January. This increased level of supercooling was maintained throughout winter until feeding commenced on the new growth of beech leaves in spring. A similar pattern of changes in supercooling, but with a much greater summer to winter difference was observed with *R. inquisitor*, with supercooling points of newly emerged beetles ranging from − 7.5 to − 13 °C in September and October, compared with − 25 to − 30 °C from November to March (Zachariassen 1985). Many studies on seasonal variations in cold-hardiness have been based on samples of insects collected between autumn and the following spring and Zachariassen (1985) has commented that, under such a programme, the first samples may be taken too late to show the initial phase of winter hardening. Danks (1978), arguing for a more ecological approach to research on cold-hardiness, noted that few studies were based on complete life cycles and thus it was difficult to interpret the selection forces encountered by insects from different winter and summer stresses.

The seasonal change of increased winter supercooling observed in many freezing intolerant insects is derived from a combination of factors that start to act in late summer and continue through to midwinter. The most important of these factors are: (i) the removal or inactivation of haemolymph or cellular nucleators; (ii) cessation of feeding and evacuation of gut contents; (iii) synthesis of cryoprotectant (polyol) antifreezes; (iv) avoidance of contact with body surface moisture; and (v) regulation of body water content. (The action of antifreeze (thermal hysteresis) proteins is also vitally important to some freezing intolerant insects and is discussed on page 109.) The importance of any one of these factors will vary greatly with different species and many studies have selected only one or two factors, usually feeding and cryoprotectants, for detailed investigation. When considering this range of factors it is important therefore to remember that first, in nature they act together whereas laboratory research tries to isolate each factor to determine its relative importance, and secondly, there is a limited amount of information on some factors, such as the action of nucleators outside of the digestive system and inoculative freezing from body surface moisture.

Nucleators

Most literature on insect cold-hardiness distinguishes between the nucleating activity of gut contents (discussed on p. 114) and nucleators found in the cells and haemolymph. There have been comparatively few studies on intracellular nucleators, particularly in freezing intolerant insects, and this remains one of the most under-researched areas of insect cold-hardiness.

If an insect is to survive in winter by a freeze avoidance strategy it must remove or inactivate all potential nucleators and those found in the haemolymph or cells may present a greater problem than food particles in the gut, which can, in theory, be evacuated from the digestive tract.

The risks to freezing intolerant insects of intracellular nucleators can only be studied in the absence of complicating factors such as feeding. Zachariassen (1982) was able to demonstrate a 'reactivation' of nucleators when a freezing intolerant insect was 'warm-acclimated' from its winter-hardened state (Fig. 4.8). When adult *Bolitophagus reticulatus* (Coleoptera: Tenebrionidae) were kept at 20 °C, supercooling points increased from -25 to -20 °C in 3 days (with a concomitant decrease in the haemolymph glycerol concentration) and then rose abruptly to -8 °C after 4 days and remained steady at that level. Earlier studies by Zachariassen (1980) had demonstrated the absence of nucleators from the haemolymph of active freezing intolerant insects, and thus it was argued

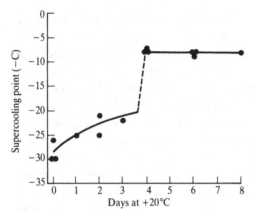

Fig. 4.8. Supercooling points of *Bolitophagus reticulatus* (starved) during warm acclimation at 20 °C (after Zachariassen 1982).

that intracellular compartments were the most probable site for the nucleating agents. Proteins and carbohydrates can act as nucleators for the freezing of water (Duman and Patterson 1978; Krog *et al.* 1979) and cells contain a wide variety of both types of compounds.

If protein nucleators in cells are to be eliminated prior to winter, Zachariassen (1985) suggested that some 'specific protein catabolic mechanism' would be required. At present there seems to be little evidence for such a mechanism, although the concentration of the amino acid alanine increased greatly in larvae of *Nemapogon personellus* (Lepidoptera: Tineidae) when maintained at low temperature or following anoxia (Sømme 1967) and decreased markedly in adult *R. inquisitor* when transferred from conditions of low temperature and anoxia to air at 20 °C (Zachariassen and Pasche 1976). Other amino acids in *R. inquisitor* (glutamate, proline and lysine) were not affected by the treatment and protein catabolism therefore seems to be a highly specific process. However, it is known that alanine and other amino acids increase in insects during cold acclimation (Storey 1984), so that this change may be unrelated to any attempt by the insect to catabolise protein nucleators.

An alternative mechanism for the winter removal of inactivation nucleators has been described in *R. inquisitor* by Baust and Zachariassen (1983). The haemolymph of the beetle is nucleator-free throughout the year but mean supercooling points range from −7 °C in summer to −24 °C in winter. Homogenisation of cold-hardy beetles that had a high capacity for supercooling produced a homogenate that gave a high nucleator activity associated with membrane structures. As the nucleators were present in winter (i.e. at a time when the beetles show extensive supercooling) it was concluded that the nucleators were inactivated by sequestration within the lipid phase of the cell membrane.

Cryoprotectants in freezing intolerant insects

Freezing intolerant insects have to avoid freezing to survive and for this reason it is generally accepted that cryoprotectants in such species function primarily as antifreeze agents. It is now recognised that there are two main groups of cryoprotectant antifreezes in freezing intolerant insects. (i) Low molecular weight polyhydroxy alcohols (polyols) and sugars, which depress the supercooling point. (ii) High molecular weight antifreeze proteins (thermal hysteresis proteins), which depress the haemo-lymph freezing point relative to its melting point and may act to stabilise the supercooled state (Duman *et al.* 1982; Zachariassen and Husby 1982). The same polyols (particularly glycerol) are common to both freezing tolerant and intolerant insects and some freezing tolerant insects also contain antifreeze proteins in winter (Duman 1982). Both groups of cryoprotectants may also have functions other than freezing prevention in intolerant species and inhibition of freezing damage in tolerant insects. Although the majority of research on cryoprotectants has focussed on polyols, Zachariassen (1985) suggests that antifreeze proteins may be more characteristic of overwintering freezing intolerant insects than polyols such as glycerol.

Polyols and sugars   Extensive supercooling to temperatures below those likely to be encountered in the overwintering site is an essential require-ment for survival in freezing intolerant insects. This level of supercooling is usually achieved as a two-stage process: first, by the removal or inactivation of any ice nucleating agents from the haemolymph, or gut contents from the digestive system, which alone may permit supercooling to $-20\,°C$ and below, and secondly, by the prewinter accumulation of polyols and sugars that extend the inherent ability to supercool.

Following the initial discovery of glycerol in haemolymph from pupae of the silk moth, *Hyalophora cecropia* (Lepidoptera: Saturniidae), by Wyatt and Kalf (1957, 1958), Chino (1957) found that glycogen was converted into glycerol and sorbitol in diapausing eggs of the silkworm, *Bombyx mori* (Lepidoptera: Bombycidae). At the same time, Salt (1957, 1959, 1961) provided the first information on the role of polyols such as glycerol in the cold-hardiness of overwintering insects. A number of different polyols and sugars have since been found in overwintering stages of freezing intolerant insects. From a comprehensive survey of the literature covering supercooling points and cryoprotectants in over-wintering eggs, larvae, pupae and adults of different species of terrestrial arthropods, Sømme (1982) reported glycerol to be positively identified in 64 of 96 species that were screened for polyols and sugars and expressed the view that glycerol was likely to be found in most insects with a great capacity to supercool. It has been suggested that the prevalence of glycerol in cold-hardy insects was attributable to the selectivity of the biochemical analyses applied (Danks 1978) but it is evident from many recent studies (Lee *et al.* 1986) utilising GLC and HPLC systems, which can identify the full range of carbohydrates in individual insects (by comparison of retention times against known standards), that glycerol is the most common and abundant cryoprotectant. Other polyols of importance in insect cryoprotection include sorbitol (Sømme 1965a, 1967; Morrisey and Baust 1976), mannitol (Sømme 1969) and, interestingly, ethylene glycol (Gehrken 1984). Among the sugars, glucose, trehalose and fructose have been recorded in freezing intolerant insects (Tanno 1964; Sømme 1969; Block and Zettel 1980). Accurate and sensitive modern chromatographic methods have revealed a complex array of other polyols in some species (e.g. ribitol, arabitol, xylitol, inositol, rhamnitol and fucitol in juveniles of the Antarctic mite, *Alaskozetes antarcticus*; Young and Block 1980), but Sømme (1982) and Zachariassen (1985) have commented that, in some cases, the concentrations are so low that their importance in cold-hardening mechanisms is not yet established.

A number of overwintering insects possess a multicomponent cryo-protectant system, such as glycerol, sorbitol and trehalose in the freezing tolerant larvae of *Eurosta solidaginis* (Morrisey and Baust 1976), glycerol, sorbitol, glucose and trehalose in larvae of *Scolytus ratzeburgi* (Coleoptera: Scolytidae) (Ring 1977) and glycerol, mannitol and trehalose in *Crypto-pygus antarcticus* (Collembola: Isotomidae) (Sømme and Block 1982). These multifactorial systems have been considered advantageous because they prevent or reduce the possible toxic effects associated with the relatively high concentration of single components that would be required to produce the same level of cryoprotection (Baust 1973; Morrisey and Baust 1976; Ring 1977). An alternative explanation for the functioning of multicomponent systems has been offered by Zachariassen (1985): as the most commonly occurring polyol – glycerol – has no effect on enzyme activity, even at the multimolal concentrations found in some overwinter-ing insects, its primary role may be to enhance cryoprotection by increasing supercooling capacity, whereas other polyols and sugars, accumulated to lower concentrations may reduce enzyme activities to promote energy conservation during prolonged overwintering.

A general pattern of seasonal changes in cryoprotectants and supercooling has been observed in many species, although there may be different environmental triggers (mainly temperature but also photoperiod and moisture, see p. 129) for particular polyols or sugars in the same species. Cryoprotectant concentration usually increases from autumn onwards, with corresponding decreases in the supercooling point, i.e. increased cold-hardiness. Some time in late winter or spring the polyol and sugar concentrations decrease and the supercooling point rises. This pattern of change is well exemplified by larvae of *Retinia resinella* (Lepidoptera: Tortricidae) in Figure 4.9 (Hansen 1973) and similar inverse relationships between glycerol concentration and supercooling points have been demonstrated for larvae of *Yponomeuta evonymellus* (Lepidoptera: Yponomeutidae) and *Cydia strobilella* (Lepidoptera: Tortricidae) (Sømme 1965b). While it is assumed that, in nature, cold-hardening occurs over a period of months from summer through autumn to winter, it is now evident that increased cold-hardiness correlated with rising levels of cryoprotectants can occur after very brief exposures (2 hours) to low temperatures (see pp. 140–141).

Many of the species on which cold-hardiness studies have been made

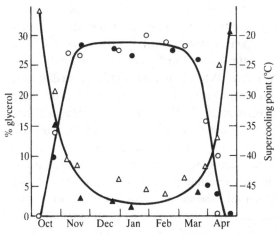

Fig. 4.9. Seasonal changes in glycerol content (percent of fresh weight) and supercooling point in larvae of *Retinia (Petrova) resinella*. ●, ▲ second instar; ○, △, last instar (after Hansen 1973).

overwinter in diapause. According to Sømme (1982) the interaction between cryoprotective substances and diapausing/non-diapausing over-wintering insects can be classified into three main types:

1. Non-diapausing species with temperature-dependent polyol synthesis, e.g. larvae of *Scolytus ratzeburgi* (Ring 1977).
2. Diapausing species that produce polyols independent of temperature, e.g. eggs of *Bombyx mori* (Chino 1957).
3. Diapausing species which accumulate polyols only on exposure to low temperature, e.g. *Limenitis archippus* (Lepidoptera: Nymphalidae) (Frankos and Platt 1976), or produce larger amounts of polyols when exposed to low temperature, such as pupae of *Hyalophora cecropia* (Ziegler and Wyatt 1975).

Our understanding of the interrelationships between diapause and cold tolerance is limited because it has been assumed that the induction of diapause and the synthesis of cryoprotectants occur independently, and usually in response to different environmental cues – photoperiod and temperature, respectively. For this reason, few studies on cold tolerance have included any comparisons between diapause and non-diapause individuals of the life cycle stage in which the overwintering diapause occurs, overlooking the role of the diapause state *alone* in enhancing winter cold-hardiness. For instance, the cabbage-white butterfly, *Pieris brassicae* (Lepidoptera: Pieridae), overwinters as a diapausing pupa. The diapause is induced by a short photoperiod acting on the fourth and fifth instar larva. When larvae are kept in either a short (diapause-inducing) or long (diapause-avoiding) photoperiod at 20 °C, a temperature that would not be expected to trigger cryoprotectant synthesis, the resultant diapause pupae contain $47 \times$ the level of sorbitol (the main cryoprotectant) than the non-diapause pupae. When maintained at 2 °C, the diapause pupae show a further increase in sorbitol, but only to $60 \times$ the level observed in non-diapause pupae at 20 °C. The role of a reliable *predictive* cue, such as photoperiod, and dynamic *indicative* cues such as temperature, in insect cold-hardiness is discussed on pages 129–137. In the light of previous comments on supercooling, cryoprotectant concentrations and the validity of using the supercooling point as indicator of the lower lethal limit of freezing intolerant insects, it is interesting to note that diapausing pupae of *P. brassicae* reared at 2 °C and containing $60 \times$ the concentration of sorbitol, supercool only 3 °C lower (to -24 °C) than non-diapause pupae at 20 °C. While both diapause and non-diapause pupae are still alive after 2 weeks at -5 °C, none of the non-diapause pupae are able to complete their development and emerge as adults on transfer to 20 °C.

The loss of polyols, such as glycerol, in spring has been linked in many species to the termination of diapause and has been observed to decrease rapidly at 20 °C, e.g. *Retinia resinella* (Hansen 1973). The nature of the relationship between the termination of diapause and the concomitant loss of glycerol may need to be re-examined in the light of current ideas on diapause (Tauber and Tauber 1976a). Historically, winter diapause has been regarded as a sequence of phases: (i) initiation by a specific environmental cue, usually decreasing photoperiod; (ii) maintenance through a period of chilling required for diapause development; (iii) termination in response to an environmental cue (increasing day length, temperature); and (iv) postdiapause development. Under this system, diapause is terminated in spring when conditions are more or less immediately favourable for growth and development. It is now apparent that, for some species, diapause ends in midwinter, at the end of the period of diapause development but with no precise termination stimulus. It is not evident when the diapause has ended because postdiapause development does not occur until some weeks or months later, when temperatures rise above the threshold for normal growth and development. In these species it is important that antifreeze cryoprotectants are lost not at the end of the diapause but only in response to warm acclimation in spring; certainly, non-diapausing species retain their cryoprotectants until the end of winter. It seems likely that the general importance of diapause termination in the spring loss of cryoprotectants has been overestimated and that the elevated temperatures that occur at the same time may be the more dominant influence.

The supercooling capacity of a system is defined as the temperature difference between its melting point and supercooling point. If the synthesis of polyols in winter depresses the supercooling point and melting point equally, although the supercooling point of the insect will be lower, the supercooling capacity will be the same. In fact many studies have now shown that the increased solute concentration of freezing intolerant insects in winter depresses the supercooling point by slightly more than twice the depression of the melting point, and this corresponds well with measurements made on nucleator-free solutions of glycerol and other solutes (Zachariassen 1985). The mechanism by which polyols depress the supercooling point more than the melting point is not clearly established; different ideas have emerged on this subject and they should not be regarded as mutually exclusive. For instance Salt (1961) suggested that the depressive effect of glycerol on supercooling points could be attributed to the increased viscosity of the solution at low temperature, which inhibits the activity of ice nucleating agents. More recently, Baust and Lee (1981) have introdced the idea of 'hydroxyl equivalents', relating to the number of hydroxyl groups in solution, which are known to affect the supercooling depression in systems that are free of nucleating agents (MacKenzie 1977) and may act in a non-colligative way, i.e. in a manner unrelated to the number of particles in solution, in cold-hardy freezing intolerant species.

A particular difficulty of relating supercooling point depressions to polyol concentrations has been highlighted by Zachariassen (1985). Many studies have expressed polyol concentration in weight units of solute per unit of fresh (wet) body weight of insect, e.g. µg glycerol per mg fresh body weight, but under this system it is not possible to compare solute types and, because the relative water content varies greatly, comparisons cannot be made between different insect species. If the weight of solute is known, together with the wet and dry weight of the insect, solute concentrations can be expressed in units of osmolality that allow comparisons between different solute types and insects and linearise the relationship between cryoprotectant concentration and supercooling points (Zachariassen 1985).

It is now widely acknowledged that polyols and sugars may have important roles in freeze avoiding insects other than their antifreeze properties. Again the evidence for some of these functions is equivocal but Ring (1980) and Duman *et al.* (1982) have summarised the possible additional functions. For instance, glycerol may stabilise enzymes at low temperature by protecting against protein denaturation (Hochachka and Somero 1973). Glycerol is an important antidesiccation agent in anhydrobiotic organisms (Crowe and Clegg 1973) and may have a similar role in overwintering insects, which characteristically have a reduced water content. Glycerol also binds water, which further acts to prevent water loss (Ring and Tesar 1981; Storey *et al.* 1981). Trehalose is one of the most common sugars in overwintering insects and, under conditions of desiccation, may be preferred to glycogen as an energy store because of its greater solubility. The reversible synthesis of glycerol from energy stores in the insect in autumn allows it to function as an energy source (directly or after metabolism) once its winter cryoprotective role has ended (Danks 1978) and Ashwood-Smith (1970) regards this as a primary function of this polyol.

Antifreeze (thermal hysteresis) proteins 'Thermal hysteresis', describes a situation in which the body fluid of an organism can be cooled below its melting point without the immediate growth of an ice crystal; the temperature at which ice crystal growth begins in such a system is the hysteresis freezing point (Zachariassen 1985) and the extent of thermal hysteresis is measured as the temperature difference between the melting point and the hysteresis freezing point of a volume of haemolymph. The agents that produce thermal hysteresis are proteins and glycoproteins and were first described by Ramsay (1964) in studies on the beetle *Tenebrio molitor* (Coleoptera: Tenebrionidae), although at the time it was thought that the compounds were associated with water balance.

The importance of thermal hysteresis as an antifreeze phenomenon was recognised by DeVries in research on some species of polar teleost fish (DeVries 1982). The melting point of the blood plasma of the fish is less than $-1\,°C$, whereas the temperature of their sea water habitat is about $-2\,°C$; fish are unable to regulate their body temperature and therefore the polar species live in equilibrium with their external environment at around $-2\,°C$. DeVries was able to demonstrate the occurrence in the body fluid of macromolecules (proteins or glycoproteins) that did not affect the melting point but that prevented the growth of ice crystals until approximately $-2.5\,°C$. Although this represents a thermal hysteresis of only $1-2\,°C$, the solute concentration of sea water is such that the temperature of the water never falls to $-2.5\,°C$ and the fish are thus protected from internal nucleation or external inoculation of freezing.

More recently, thermal hysteresis proteins have been found in an increasing number of insect species and, in recognition of their envisaged primary role as antifreeze agents are now usually described as antifreeze proteins. The occurrence of antifreeze proteins in an overwintering insect was first reported by Duman in the freezing intolerant larvae of the tenebrionid beetle *Meracantha contracta* (Coleoptera: Tenebrionidae) (Duman 1977a,b). The larvae overwinter in decomposing logs that in turn may be covered by an insulating layer of snow. Interestingly, the larvae do not accumulate polyols but show a seasonal increase in antifreeze proteins that are absent in summer (Fig. 4.10 and Table 4.3). Thus, the maximum hysteresis of $3.71\,°C$ is seen in midwinter (Fig. 4.10) and is accompanied by a summer to winter depression of the supercooling point from -5 to $-11\,°C$ (Table 4.3). While the influence of antifreeze proteins on supercooling is far from clear, there is no doubt that the winter depression of the haemolymph freezing point of *Meracantha* to $-5\,°C$ lessens the possibility of inoculative freezing across the cuticle from the

Table 4.3. *Haemolymph melting point, freezing point and thermal hysteresis and whole-body supercooling points (all °C) of* Meracantha contracta *at various times of years (after Duman 1977b)*

Time of year	Melting point	Freezing point	Thermal hysteresis	Supercooling point
Winter	−1.31	−5.02	3.71	−11
Summer	−0.81	−0.81	0	−4

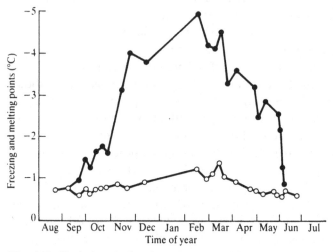

Fig. 4.10. Variations in freezing (●) and melting (○) points showing the amount of thermal hysteresis in the haemolymph of *Meracantha contracta* larvae from northern Indiana over the course of a year (after Duman 1977b).

ice that forms in and around the overwintering site; the discovery of antifreeze proteins in insects that overwinter close to external ice led Zachariassen and Husby (1982) to suggest that protection against the seeding of ice through the body wall may be an important function of this particular group of cryoprotectants.

The tenebrionid beetle *Uloma impressa* (Coleoptera: Tenebrionidae) is also freezing intolerant and overwinters as adults under the bark of old and dead trees. This species shows a low level of hysteresis in summer (1.14 °C) that increases four-fold to 4.76 °C in midwinter (when the melting point is also depressed from −0.88 °C in June to −9.9 °C in February) accompanied by high levels of glycerol and a supercooling point shift from −6.2 °C in summer to −21.5 °C in winter (Duman 1979).

The mechanism by which antifreeze proteins inhibit ice growth has not been studied in insects and available information is based on work with the polar fish. In simple terms, the antifreeze proteins are able to adsorb on to the surface of ice crystals and alter the structure of the crystal such that it is unable to act as a seed for nucleation until the temperature of the system is lowered to the freezing point, so producing the hysteretic effect. A detailed description of these processes is given in Raymond and DeVries (1977) and DeVries (1980).

Purification and identification of the components of insect antifreeze proteins is still at an early stage but Duman (1982) has made some preliminary observations and noted some important differences compared to the polar fish. For instance, all the insect antifreeze proteins lack a carbohydrate component, although glycoprotein antifreezes are common in fish; additionally the fish antifreezes may contain more than 60 per cent alanine but in insects this amino acid occurs at much lower levels. The history of insect antifreeze proteins is still young and owes much to the innovative approaches of Duman and Zachariassen, who have attempted to explain some of the apparent anomalies of these compounds and to provide answers to difficult questions. For instance, the majority of species with antifreeze proteins have hysteresis freezing points between -5 and $-10\,°C$ (Table 4.4) and, because a number of the same species may experience winter temperatures below $-10\,°C$ and also have supercooling points below $-20\,°C$ (when they would be in a highly supercooled state), Zachariassen (1985) has considered the possible roles of antifreeze proteins below the hysteresis freezing temperature. Duman (1982) has suggested that the ability of antifreeze proteins to bind and 'poison' potential seed crystals may extend to embryo crystals, so preventing them from reaching the critical size necessary to induce spontaneous nucleation in a supercooled liquid and, in turn, lowering the supercooling point. Experimental evidence for this type of effect was provided by Zachariassen and Husby (1982) with the freezing intolerant beetle *Rhagium inquisitor*. To measure thermal hysteresis, a sample of haemolymph (usually a nanolitre volume) is 'flash-frozen', warmed slowly and viewed under a microscope until a single ice crystal is held at its smallest visible size. The temperature is then gradually decreased (on a scale of milliosmoles, which can be transformed to a temperature by a simple equation) until the crystal suddenly increases in size and nucleates the sample. The level of thermal hysteresis is measured as the difference between the temperature at which the last ice crystal is held at its smallest size (melting point) and the temperature of rapid crystal growth and

Table 4.4. *Haemolymph melting point, hysteresis freezing point and whole body supercooling point (all °C) of freeze tolerant and intolerant insects with antifreeze proteins (after Zachariassen 1985)*

Species	Lowest MP	Lowest HFP	Lowest SCP	Tolerant or intolerant
Collembola				
Entomobrya nivalis	−1.0	−3.0	−18.0	INT
Coleoptera				
Platycerus caprea	−0.9	−4.7	−21.5	INT
*Dendroides canadensis**	−5.5	−11.1	−12.0	T
	−4.1	−6.5	−26.1	INT
Coccinella novemnotata	−2.1	−4.6	−13.0	INT
Cucujus clavipes	−2.4	−6.2	−9.4	T
Melanotus sp.	−5.6	−6.5	−14.0	T
Pytho depressus	−1.9	−1.9	−6.0	T
Xylita laevigata	−0.8	−2.6	−24.0	INT
Meracantha contracta	−1.3	−5.0	−11.0	INT
Eleodes blanchardi	−0.9	−0.9	−6.3	T
Tenebrio molitor	−1.3	−3.6	−15.0	INT
Uloma impressa	−9.9	−14.7	−21.0	INT
Uloma peroudi	−1.0	−8.5	−18.0	INT
Iphthimus laevissimus	−0.8	−3.5	−15.9	INT
Rhagium inquisitor	−4.3	−9.5	−27.0	INT
Rhagium mordax	−0.8	−6.7	−11.5	INT
Eremotes ater	−0.9	−6.3	−22.0	INT
Ips acuminatus	−6.3	−9.5	−31.6	INT
Mecoptera				
Boreus westwoodi	−1.2	−6.4	−8.6	INT
Orthoptera				
Parcoblatta pennsylvanica	−1.4	−3.2	−8.3	T

MP, melting point; HFP, hysteresis freezing point; SCP, supercooling point; T, tolerant; INT, intolerant.
* This species can change its overwintering strategy (tolerant or intolerant larvae) in successive years.

nucleation of the sample. Based on this experimental protocol the lowest haemolymph melting and freezing points of *R. inquisitor* were −4.3 and −9.5 °C (5.2 °C of hysteresis) and the lowest supercooling point was −27 °C. Zachariassen and Husby (1982) found a linear relationship between the hysteresis freezing point and the logarithm of crystal size and concluded that the hysteresis freezing point depended on the size of the seeding crystal, with smaller crystals producing lower hysteresis freezing points. Using the available data describing the relationship between ice crystal size and the hysteresis freezing point and extrapolating to crystal sizes below those visible under experimental conditions, Zachariassen and Husby found that at the size of the water molecule aggregates required for nucleation at $-25\,°C$ (the observed supercooling point of the beetle from which the haemolymph was taken) the hysteresis freezing point was depressed to $-25\,°C$. While this is the only study of its type, and extrapolation over such a wide range should be viewed with caution, there is nevertheless the possibility that antifreeze proteins may have the ability to stabilise the supercooled state over the entire supercooling range of a species.

It has been suggested that increased supercooling in insects that show a seasonal increase in antifreeze proteins (sometimes without any accumulation of polyols) may be partly attributable to the ability of the proteins to inactivate ice nucleating agents (Duman and Horwath 1983). The addition of a purified antifreeze protein from *Tenebrio molitor* sufficient to produce a thermal hysteresis of $0.35\,°C$, to the haemolymph of *Periplaneta americana* (Dictyoptera: Blattidae) lowered the haemolymph supercooling point of the latter by $2\,°C$ from -14 to $-16\,°C$, but the effect was lost if the sample was heat-treated to denature the protein (Duman *et al.* 1982). However, antifreeze proteins from the freezing intolerant beetle *Iphthimus laevissimus* (Coleoptera: Tenebrionidae) did not reduce the nucleator efficiency of nucleators from the freeze tolerant beetle *Eleodes blanchardi* (Zachariassen and Hammel 1976). From the limited amount of information so far available it appears that antifreeze proteins may enhance supercooling in some species but that this effect is not based on any ability to inactivate ice nucleating agents.

While the depression of the freezing temperature of the haemolymph is clearly advantageous to insects that have to avoid freezing to survive, it is not immediately obvious what role antifreeze proteins may have in species that are freezing tolerant and produce ice nucleating agents to limit their supercooling and ensure that freezing occurs at an elevated subzero temperature. This apparent anomaly was highlighted by Duman (1982), who suggested that antifreeze proteins may be 'multifunctional' in the same way that glycerol can increase supercooling in freezing intolerant insects, cryoprotect the tissues of freezing tolerant species and contribute to desiccation resistance in both groups. Possible roles for antifreeze proteins in freeze tolerant insects are discussed on page 94 and in the section on environmental triggers for different cold-hardiness agents (see p. 135).

In concluding this section it is interesting to note Zachariassen's (1985) suggestion that antifreeze proteins may be more characteristic than polyols of freezing intolerant insects. There appear to be advantages for the antifreeze effect in insects to be based on proteins rather than low molecular weight polyols and carbohydrates, as recognised by Duman *et al.* (1982). For instance, antifreeze proteins function in a non-colligative manner and so do not produce the large increases in osmotic pressure associated with high concentrations of polyols; additionally carbohydrate and polyol antifreezes, such as glycerol and trehalose, occur in important metabolic pathways and thus may require the adjustment of complex control mechanisms in the autumn synthesis and spring catabolism of these compounds.

Feeding (gut contents)

The effect of gut contents from feeding as a factor that reduces supercooling ability is one of the most clear-cut relationships in insect cold-hardiness (Sømme 1982) but as Sømme (1985) and Baust and Rojas (1985) commented, the actual site of nucleation has rarely been observed and there are many anomalies in what is essentially a correlation between gut content and supercooling capacity.

General theory attributes the reduced supercooling ability of fed or feeding insects to the initiation of freezing by nucleating agents in food residues. Sømme (1982) cites numerous studies in which there is a marked difference in supercooling in insects depending on their feeding status. For instance, the mean supercooling point of larvae of *Agrotis orthogonia* (Lepidoptera: Noctuidae) decreased from $-10.3\,^{\circ}\mathrm{C}$ when feeding on wheat to $-23.6\,^{\circ}\mathrm{C}$ in the premoult condition when the gut contents are evacuated (Salt 1953). Similarly, the mean supercooling point of larvae of the Mediterranean flour moth, *Ephestia kuhniella* (Lepidoptera: Phycitidae), changed from $-11\,^{\circ}\mathrm{C}$ when feeding on whole wheat flour to $-20.5\,^{\circ}\mathrm{C}$ in the premoult state. However, neither species showed any increase in supercooling when starved, although, as Sømme (1982) observed, *Agrotis* and *Ephestia* (and a number of other species that show no supercooling response after starvation) do not experience extreme cold in their natural environment and in some cases experiments were conducted on non-overwintering stages that would not survive or respond to conditions of starvation.

In nature, most overwintering insects of the temperate or colder climates evacuate the digestive tract as part of a multifactorial process that maximises supercooling in winter. The particular importance of the starved condition (or one in which any food particles remaining in the gut are prevented from acting as nucleators) is clearly demonstrated by overwintering adult beech weevils, *Rhynchaenus fagi*, where supercooling to −22 °C is retained through winter to spring but rises to −16.8 °C as soon as the weevils feed on the emerging beech leaves (Bale 1980).

The influence of feeding and gut contents on supercooling has been well studied in a number of Collembola and mite species. Sømme and Conradi-Larsen (1977) recorded supercooling points and examined the gut contents of *Tetracanthella wahlgreni* (Collembola: Isotomidae). There was a clear demarcation of supercooling points at −15 °C when individuals with 'gut contents' were above and those with 'empty guts' were below −15 °C. A marked bimodal distribution of supercooling points into high (usually above −10 °C) and low (usually below −20 °C) groups has been demonstrated in the Antarctic collembolan *Cryptopygus antarcticus* and the mite *Alaskozetes antarcticus*, as shown in Figure 4.11 (Block 1982b; Sømme and Block 1982). In general, insects collected in the summer when feeding occurs had supercooling points (SCP) consistently in the high groups whereas winter populations were strongly aggregated in the low group. Ninety-two per cent of a summer sample of *Cryptopygus* were in the high group, with a mean SCP of −7.2 °C. Following starvation at 5 °C for 6 days there was a distinct shift in supercooling, with 56 per cent in the low group with a mean SCP of −24.8 °C. Sømme and Block (1982) were also able to show that dietary components differed in their effect on supercooling. Thus, 87 per cent of *Cryptopygus* fed on moss turf homogenate formed a high group with a mean SCP of −6.7 °C, whereas a diet of purified green algae produced a majority (65 per cent) in the low group, with a mean of −24.1 °C.

A notable feature of the Antarctic insects was the consistency of supercooling with season, particularly in the low group in winter. A different pattern has been observed with the temperate collembolan *Hypogastrura* spp. (Collembola: Hypogastruridae) where, at intervals in winter and depending upon the prevailing weather, the majority of the population may be in the high group or low group (around −7 and −25 °C, respectively), evenly split between the groups, or moving to the low group via a mid-group between −12 and −15 °C (Bale and Pullin 1991). Following a prolonged cold spell with some severe frosts, the majority of the population is in the low group; within days of a mild spell in midwinter there is a sharp reversal and supercooling points are almost exclusively in the high group. Experimental starvation and feeding of *Hypogastrura* on moss and lichen showed a very strong correlation between feeding and supercooling but also revealed a distinct midgroup

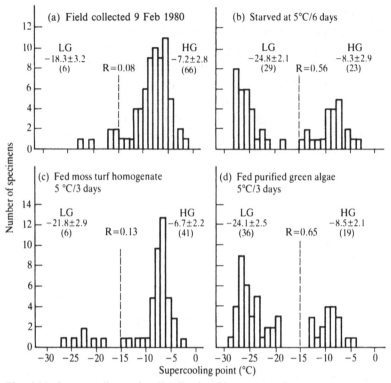

Fig. 4.11. Supercooling point distribution histograms of *Cryptopygus antarcticus*. (a) field collected specimens, (b) specimens starved at 5 °C, (c) fed on moss turf homogenate, (d) fed on purified green algae. LG, low group supercooling point; HG, high group supercooling point; stippled line indicates division between low and high groups; R indicates proportion of each sample in the low group. Supercooling points ± standard error are presented for each low and high group (after Sømme and Block 1982).

that formed within a few days of entering a starvation regime; the low group status was only attained after 10–14 days without food. It has been suggested that as digestion proceeds in a starved insect the food within the gut becomes concentrated in parts of the digestive tract where it is less potent as a nucleator source. In ecological terms *Hypogastrura* is clearly an opportunist, feeding in winter whenever conditions are favourable but running a clear risk of a freezing death when mild weather changes suddenly to severe cold, allowing little or no time for a starvation-mediated increase in supercooling. However, if feeding during these intermittent periods of higher temperatures is sufficient for development and moulting, the removal of the gut linings as part of ecdysis may remove any food nucleators and thus ensure that a proportion of the population will always be able to supercool to below $-20\,°C$.

Insects with special feeding mechanisms such as 'sap sucking' aphids are little affected by the imbibition of phloem sap; although there is a slight loss in supercooling in new born nymphs after their sap feed (Powell 1976; O'Doherty and Bale 1985). In general, aphids feeding on healthy host plants invariably supercool to below $-20\,°C$ (O'Doherty and Bale 1985; Knight and Bale 1986). An interesting exception is the black bean aphid, *Aphis fabae* (Hemiptera: Aphididae), which supercools to around $-24\,°C$ when feeding on the summer host – field beans, *Vicia fabae* – but shows a dramatic change to -10 to $-13\,°C$ when transferred to the winter host – spindle, *Euonymus europaeus* (Gash and Bale 1985; O'Doherty 1986). It seems that spindle sap contains a nucleating agent that is active soon after aphids feed on the leaves but is only slowly lost from the aphid digestive tract after the reverse transfer to a bean host.

With the exception of these 'specialist feeders', there is evidence that the partly-digested food of insects that feed on fresh leaves may be 'freezable' in the gut or contain indigenous nucleators. Although dry food particles such as flour and other abiotic agents are not themselves freezable, they may act as nucleating sites within the semiliquid medium of the digestive tract (Salt 1953; Sømme 1982). Additionally, the food of most insects, whether feeding on leaves or stored products (grain, flour) is likely to be contaminated with atmospheric dust particles that can act as efficient nucleators when passing through the gut system (Salt 1961). There is certainly a case to be made for the proposition that the gut is a 'probable prime site' for nucleation in freezing intolerant insects. Baust and Rojas (1985) have drawn together the experimental evidence that supports this view. Thus, most insects 'clear the gut' in preparation for winter; diets that are rich or enriched with nucleators can modify supercooling; starvation can enhance supercooling; newly emerged insects with an empty gut may show a reduction in supercooling after their first feed; and the bimodal (or trimodal) distribution of supercooling points (as in the Antarctic Collembola and mites (Block 1982b; Sømme and Block 1982)) can frequently be correlated with full, clearing or empty guts. As a counterbalance to these experimental results, the same authors provide an even longer list of contradictions and exceptions, again based on experimental observations rather than supposition:

1. Many freezing intolerant species do not evacuate their gut in preparation for winter.

2. Manipulation of the nucleator content of the diet has only a short-lived effect, or no effect at all, on supercooling.
3. Starvation may not enhance supercooling (although this may arise because the gut contents are retained (Parish and Bale 1990)).
4. Supercooling of newly emerged insects may not be affected by feeding (but note that the example is a phloem-feeding aphid in which the diet is primarily sugars and polyols and likely to be nucleator free).
5. Insects contain potential nucleators in other body compartments (haemolymph and other tissues).
6. Body surface moisture can reduce supercooling dramatically.
7. Body water content can influence supercooling.
8. Piercing, sucking and chewing insects may feed on highly active plant nucleators (such as the ice-nucleating bacteria *Pseudomonas syringae*) without postfeeding effects on supercooling, although in the case of sap feeders such as aphids it is not clear whether the bacteria are of sufficiently small size to pass through the 'sap channel' formed by the stylets.

In summary, Baust and Rojas acknowledge that, in the absence of haemolymph nucleators or the effects of surface moisture, there are many correlations that suggest the gut to be the site of nucleation in freezing intolerant insects, but they argue that the formation of an ice nucleus is a molecular-based phenomenon that researchers of insect cold-hardiness have attempted to study at the whole organism level. It is perhaps wise to conclude that, in the same way that a desire to establish 'general principles' of cold-hardiness has led to an over-simplistic classification of insects as freezing tolerant or intolerant (which cannot accommodate the prefreeze species such as aphids) the recognition of the gut as the major site of nucleation (which for many species is undoubtedly correct) may have become accepted as part of a dogma that has not yet appreciated the limits of the methodology upon which it is based.

Contact moisture The internal changes that an insect makes in preparation for winter, such as the evacuation of the digestive system and synthesis of cryoprotectants to maximise supercooling can be undermined by the freezing of 'surface moisture' on the cuticle. While the insect cuticle is regarded as waterproof, on the basis of its wax layer, there are a number of orifices associated with feeding, excretion and respiration through which water and ice may penetrate from the exterior surface to internal areas. Body surface moisture can therefore cause 'inoculative freezing' of an insect at temperatures much above the inherent and internal ability to supercool.

The effect of contact moisture on supercooling in freezing intolerant insects is not well studied and the situations in which insects are at risk from this effect in nature are difficult to mimic in the laboratory. Clearly there are a number of interacting factors and this may explain the very varied response of different species to the presence of body surface moisture. For instance, the lowest winter supercooling point of 'dry' adults of the beech leaf mining weevil, *Rhynchaenus fagi*, is about $-23\,°C$. When weevils are immersed in water and then placed in narrow capillary tubes to form a film of moisture around the specimen the mean supercooling point is raised to $-9\,°C$, and all individuals were found to die from inoculative freezing between -8 and $-9\,°C$ (Bale 1980). A different pattern of results was obtained with the aphid *Myzus persicae* (Hemiptera: Aphididae) when dry specimens were cooled with a large droplet of water placed on the abdomen. In most cases the droplet froze at temperatures in the region of -6 to $-12\,°C$ but the normal aphid supercooling point was recorded below $-20\,°C$, indicating that no inoculation had occurred when the water froze (O'Doherty and Bale 1985). It seems likely that the cuticle of different species, possibly related to external structures or areas that are more easily penetrated by ice, determines the susceptibility of the insect to surface moisture effects and that this may interact with the volume and distribution of water present. Butterflies such as the small tortoiseshell, *Aglais urticae* (Lepidoptera: Nymphalidae), and peacock, *Inachis io* (Lepidoptera: Nymphalidae), show a marked reduction in supercooling when a fine mist is sprayed on to their body, and it is possible that the hairy covering of the head and thorax may retain small droplets, which then collect at the base of the body hairs (Bale and Pullin unpublished data). Insects that overwinter on cereal plants or in the bark crevices of trees may become trapped within droplets of water and encased in ice when the water freezes. For some insects the availability of 'dry' overwintering sites assumes crucial importance and the avoidance of high humidity or contact with free moisture may be a prerequisite for the realisation of the survival potential of the acquired supercooling ability.

The soil environment harbours many species of microfauna, particularly mites and Collembola, which are very sensitive to desiccation and consequently usually occupy habitats with a high or saturated humidity (Sømme 1982). The soil habitats of the Antarctic are frozen at the surface (and to a variable depth) more or less continuously during the winter and yet the main microfauna species survive in consistent numbers (Block

1982c). Either the reproductive powers of the survivors are more than adequate to compensate for the winter mortality, or the species may show a behavioural response to low temperature and migrate deeper into the soil profile, below the level of frost penetration assisted by the insulation from snow and litter cover at the surface (van der Woude and Verhoef 1986).

While the inoculation of freezing from body surface moisture is clearly a potentially lethal event in some freezing intolerant insects, the importance of external nucleation has not been considered in freeze tolerant species. It is generally recognised as advantageous for freeze tolerant insects to freeze at a high subzero temperature and therefore if surface moisture induces freezing at a higher temperature than an internal nucleator, and in an extracellular site, this effect may not be harmful to these insects. In fact freeze tolerant larvae of the gall fly, *Eurosta solidaginis*, frozen in a film of moisture show a higher survival to adult emergence than larvae frozen dry or unfrozen specimens (Bale *et al.* 1989a).

At present it seems that insects differ in their susceptibility to inoculative freezing at the species level and that this characterisic is based, at least in part, on the number of penetratable areas on the cuticle, the extent of the coverage of water on the body surface and the shelter provided by the overwintering site.

Body water content and dehydration   A decreasing body water content may be a natural phenomenon in insects preparing for winter and may contribute to the 'cold-hardening' process by increasing the concentration of solutes. Certainly, there is a dehydration effect resulting from the evacuation of gut contents. This occurs in most insects before winter and particularly in larvae that overwinter in diapause (Ring 1980). Under experimental conditions changes in cold-hardiness, as assessed by supercooling, in response to artificial dehydration, are extremely variable. Some of the earliest experiments on insect cold-hardiness by Payne (1927a,b) described a marked increased in supercooling when larvae of *Synchroa punctata* (Coleoptera: Melandryidae) were desiccated over calcium chloride. Subsequent experiments by Salt (1956) on eggs of *Melanoplus bivittatus* (Orthoptera: Acrididae) and larvae of *Cephus cinctus* (Hymenoptera: Cephidae) and Sømme (1964) on prepupae of *Diplolepis radicum* (Hymenoptera: Cynipidae) found that dehydration had to be severe, and in some cases life-threatening, to produce appreciable changes in cold-hardiness. The body water content of *Pytho deplanatus*, which overwinters

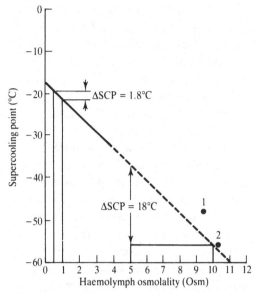

Fig. 4.12. Effect of a two-fold increase in solute concentration on supercooling points in insects with high and low body fluid osmolality. The line is based on observed supercooling points of freezing intolerant insects to about $-30\,^{\circ}$C and extrapolated (dashed line) to haemolymph osmolality values of insects with very low supercooling points. Observed values for (1) *Bracon cephi* and (2) *Rhabdophaga strobiloides* are shown (after Zachariassen 1985).

as freeze tolerant larvae under the bark of fallen spruce trees, was reduced from 69 to 30 per cent during a cold acclimation regime, this was accompanied by an increase in supercooling ability from -4.6 to $-54\,^{\circ}$C (the lowest recorded supercooling point for a freeze tolerant species). Body water content remained consistent in two other freeze tolerant larvae occupying the same niche, *P. americanus* and a *Xylophagus* species, throughout a prolonged acclimation period and the supercooling points remained high, -3.5 to $-6.6\,^{\circ}$C, typical values for freeze tolerant larvae with effective ice nucleators.

An explanation for the varied effect of changes in body water content on supercooling capacity has been provided by Zachariassen (1985) as described in Figure 4.12. In insects that avoid freezing by supercooling, a lowering of the body water content increases their cold-hardiness by increasing the concentration of cryoprotectants and lowering the supercooling point. If the body water content of a warm-acclimated insect with a normal body fluid osmolality of 500 mosM/kg was reduced by 50 per cent, the osmolality would increase to 1000 mosM/kg and the supercooling point would be lowered by 1.8 °C. If, on the other hand, the body water content of a cold acclimated insect that had accumulated cryoprotectants to a level of 5000 mosM/kg was reduced by 50 per cent, the osmolality would be raised to 10 000 mosM/kg and the supercooling point would be lowered by 18 °C; the effect of a given reduction in body water content on supercooling depends therefore on the osmolality of the body fluids, and the lowering of supercooling points will be greatest in insects with high solute concentrations (Zachariassen 1985).

In general, the loss of a proportion of the body water in autumn is a favourable adaptation that contributes to a seasonal increase in cold-hardiness. However, there must be an optimal 'trade-off' between this partial dehydration, which increases supercooling and reduces the likelihood of nucleation in the digestive system, and the need for the insect to maintain its water balance within certain limits. In a comparative survey of survival in relation to water loss in Antarctic microarthropods, Worland and Block (1986) found that the mite species studied were more resistant to dehydration than the Collembola. For instance, when exposed to 5 per cent relative humidity (RH) at temperatures from −10 to +50 °C, *Parisotoma octooculata* (Collembola: Isotomidae) lost water at 20 per cent/hour at −10 to +10 °C, *Cryptopygus antarcticus* lost water at 5 per cent/hour at −10 °C, and *Archisotoma brucei* (Collembola: Isotomidae) at 8 per cent/hour at 10 °C and 25 per cent/hour at 0 °C. In contrast, the mite *Halozetes* lost less than 1 per cent/hour at temperatures from −10 to +20 °C and *Alaskozetes* only 5 per cent/hour at 35 °C (Fig. 4.13). When survival of the species was compared at 5 per cent RH and temperatures from 0 to 20 °C, no mortality occurred at any temperature until the sample had lost between 14–18 per cent of its fresh weight, and the highest mean survival occurred at 10 °C (Fig. 4.14).

The importance of the partially dehydrated state in winter, which contributes to increased supercooling and also appears to reduce the effectiveness of gut nucleators, has been demonstrated in *Cryptopygus antarcticus* (Cannon *et al.* 1985) and *Alaskozetes antarcticus* (Cannon 1986a). When the mite *Alaskozetes*, in a cold-hardy state with high glycerol levels and low supercooling points (−30 °C), was maintained at 4 °C and 100 per cent RH and given access to distilled water or 10 per cent glucose solution, there was an extreme loss of supercooling of 20–25 °C and the virtual elimination of glycerol. There was only a small elevation of supercooling points (3–4 °C) in the absence of available liquids, indicating the resistance of the mite cuticle to water vapour, although Cannon (1986b) has shown that over long time periods and in

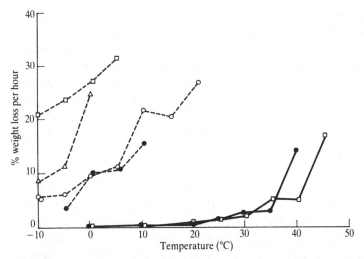

Fig. 4.13. Rates of weight loss of four species of micro-arthropods under desiccating conditions (RH = 5 ± 1 per cent) at controlled temperatures. ●—● *Alaskozetes antarcticus* adults, □—□ *Alaskozetes antarcticus* deutonymphs, ○-----○ *Cryptopygus antarcticus* mature, ●-----● *Cryptopygus antarcticus* juvenile, □-----□ *Parisotoma octooculata* mature, △-----△ *Archisotoma brucei* adults.

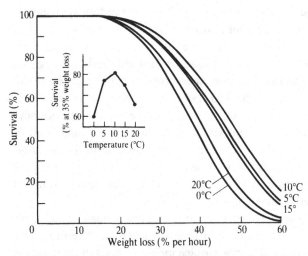

Fig. 4.14. Percentage survival of adult *Alaskozetes antarcticus* against weight loss at different temperatures. Inset: data extracted for 35 per cent weight loss showing the optimum temperature for survival (after Worland and Block 1986).

a saturated atmosphere the supercooling potential decreases to the same extent as when free water is available. A similar pattern of results was observed with *Cryptopygus*, where the availability of distilled water caused a substantial loss in supercooling associated with a marked increase in body water content. The effect of water uptake was to reduce supercooling to the level normally associated with summer feeding stages (Cannon *et al.* 1985). The uptake of free water and increased body water content apparently reactivates gut nucleators that are effectively masked during the winter dehydrated state.

Length of exposure   A supercooling experiment starting from an ambient temperature of 20 °C with an insect that freezes at -20 °C will run for 40 minutes at a cooling rate of 1 °C/minute. Leaving aside the possible effects of cooling rate on the observed supercooling point and the likelihood of surviving the freezing process, an experimental protocol that subjects individual specimens to temperatures that decrease linearly at 1 °C/minute clearly differs in its *time* component from the situation experienced in nature.

When a liquid medium is steadily cooled, a temperature where spontaneous freezing occurs is eventually reached. Nucleation, which occurs at the limit of supercooling, it a probabilistic phenomenon due to the chance orientation of water molecules into or on to an ice lattice (Baust and Rojas 1985). Homogeneous nucleation occurs in pure water by the development of an ice nucleus of critical size from the random aggregation of water molecules (Meryman 1966). Most nucleation in insects is thought to be heterogeneous – freezing being caused by the contact of water with other substances, such as particles of food or dust although Salt regards the events leading up to freezing in a supercooled insect as essentially the same as that in any liquid (Salt 1958, 1961, 1966b; Sømme 1982). While it is evident that there is considerable variation in the 'nucleator efficiency' of different substances, for any nucleator:water system, the likelihood of ice formation increases as the temperature is decreased, until the probability of nucleation approaches unity (Baust and Rojas 1985). At a selected experimental cooling rate (usually 1 °C/minute) this temperature is recorded as the supercooling point.

While it is generally accepted that an individual insect possesses a 'fixed' supercooling point at any moment in time (which is predetermined by the species, stage, nutritional status, surface moisture, level of acclimation and cooling rate (Salt 1950), two related areas first recognised by Salt and highlighted again by Sømme (1982) have received far less attention than the simple determination of supercooling points. First, since in theory the chances of nucleation increase both with decreasing temperature at a fixed cooling rate and with increasing period of exposure at a fixed temperature, do insects freeze if cooling is terminated at a subzero temperature above the supercooling point and held at that temperature for increasing time intervals? Secondly, does prolonged exposure in the supercooled state produce deleterious effects (including death) even if the insect does not freeze?

When viewed in ecological terms, the causes of the cold-induced death above the 'normal' supercooling point following prolonged exposure are not so important as the death itself, but in terms of our understanding of the physiology and biochemistry of insect cold-hardiness, identification of these mortality processes is an area of central importance. Mature larvae of the wheat stem sawfly *Cephus cinctus* supercool to between -20 and $-30\,°C$. When a sample of 20 larvae was maintained at $-20\,°C$, 12 larvae froze in the first 48 hours and the remainder froze after exposures of between 5 and 82 days. The experiment was repeated on two occasions when, in the first batch, 4 larvae were still unfrozen after 120 days while in the second group all had frozen after 26 days (Salt 1950). Apart from the obvious variations in cold-hardiness between and within the samples of insects, the larvae of *C. cinctus* showed an important feature to aid the investigator: when frozen, the larvae became white and readily distinguishable from unfrozen larvae. In 'time' exposures at subzero temperatures it is essential to be able to distinguish dead and frozen insects from those that are dead but unfrozen. Unless there is some obvious colour change, as with *C. cinctus*, the only certain method of detecting ice formation is to attach specimens to temperature sensors (thermocouples) throughout their cold exposure. A number of reports include 'mortality curves' showing that mortality increases at a fixed subzero temperature with increasing period of exposure (Sømme 1982) but it is not clear whether deaths were assumed to be by freezing, in the light of Salt's discovery with *C. cinctus*, or whether freezing was proved to be the cause of death. Hansen (1975) noted seasonal changes in the LT_{50} (time) of pupae of *Mamestra persicaria* (Lepidoptera: Noctuidae) over a period when the supercooling capacity of the insects remained constant, which may indicate that some of the deaths were not caused by freezing.

Insects that do die primarily from freezing when maintained in the supercooled state above their supercooling point, such as *C. cinctus*, do not show a linear relationship between freezing and time (Salt 1950, 1961). However, when the time periods required to produce freezing at different

constant temperatures are plotted on a logarithmic scale, the relationship
becomes linear and from the regression equation for the line (Salt 1966c)
it is possible to derive time intervals as predictions of when freezing will
occur at particular constant temperatures; for *C. cinctus* these vary from
1 second at $-30\,°C$ and 1 minute at $-27\,°C$, to approximately 1 month at
$-19\,°C$ and 1 year at $-17\,°C$. However, longer exposure periods, such
as 6–12 months are both ecologically and physiologically unreal because
the vast majority of overwintering insects do not experience such prolonged cold stress in their natural environments. Furthermore, extrapola-
tions from a regression line relating to such long periods do not take into
account the possibility that cold-induced lethal processes other than
freezing may become more important with increasing time in the supercooled state (Salt 1950).

It seems reasonable to conclude that freezing is more likely to be the
cause of death when insects are maintained at temperatures close to the
'normal' supercooling point and die after short exposures, than when kept
at temperatures 15–20 °C above the supercooling point, as determined by
a cooling rate of 1 °C/minute, and death occurs some weeks or months
later. Following the experiments with *C. cinctus*, Salt (1950) recommended
that laboratory studies on winter mortality should cover the entire
temperature range from the temperature where the time required to freeze
(kill would be a more appropriate term) all specimens is a matter of only
a few minutes up to one where at least some freezing (death) occurs during
a period of 4 or 5 months. It is also clearly desirable in these time
exposures to subject insects to fluctuating temperatures typical of those
experienced in natural environments, but this is technologically difficult
and most recent research has compared mortality in constant temperature
exposures in the laboratory with that observed in 'caged' populations
placed in typical habitats.

The goldenrod gall moth, *Epiblema scudderiana* (Lepidoptera: Olethreutidae), overwinters as mature larvae in stem galls on the goldenrod plant
in North America. Mean supercooling points of the larvae decreased from
$-13.9\,°C$ in the early autumn to stabilise between -35 and $-40\,°C$
throughout the winter. In a winter, when the lowest recorded temperature
at the study site was $-26\,°C$, there was a 90 per cent successful pupation
and emergence from a field population in spring, and 100 per cent survival
of winter larvae maintained at $-18\,°C$ for 197 days (Rickards *et al.* 1987).
Clearly this species survives by virtue of its extensive supercooling,
and there is little or no time related freezing or prefreeze mortality.
The Antarctic is another environment where there is a prolonged and

unremitting winter. The collembolan *Cryptopygus antarcticus* is a relatively abundant species in the very limited arthropod fauna of the Antarctic. The mean supercooling point of *Cryptopygus* decreases seasonally from −5 to −10 °C in summer to about −25 °C in winter, associated with the evacuation of gut contents, partial dehydration and an increase in total potential cryoprotectant concentration. In a standard test exposure to −15 °C for 24 hours, the low survival of summer field samples (less than 50 per cent) increased to more than 80 per cent in winter populations (Block 1987). However, field populations showed overwintering declines of approximately 22 per cent in a moss turf (where the density was higher) and 79 per cent in a moss carpet (Block 1982c). The reasons for these substantial winter field mortalities are unknown but may be related to time induced freezing in the supercooling state, chill injury (prefreeze), or other physiological stresses that occur with the changing season.

A much greater disparity between measured supercooling capacity and mortality occurring at less severe temperatures over longer periods of time has been observed with the bertha armyworm, *Mamestra configurata* (Lepidoptera: Noctuidae). This species overwinters as diapausing pupae 2–15 cm deep in the soil with winter supercooling points between −18 and −22 °C; although the soil temperature at 5 cm depth is consistently below 0 °C from November to March or April, the minimum temperature at that depth is −15 °C in 25 per cent of winters and −20 °C in only 5 per cent of winters. Superficially, therefore, the supercooling capacity of the pupae would appear to afford sufficient protection against a freezing death in most winters (Turnock *et al.* 1983). Diapausing pupae were exposed for 140 days to a constant −5, −10 or −15 °C, or to combinations of −5, −10, −15 and −20 °C. The main effects of these treatments were: (i) reduced survival to postdiapause stages; (ii) reduced survival to emergence (as either malformed or normal adults); and (iii) a reduced rate of postdiapause development. The results indicated that there was a continuum of effects ranging from death during the diapause stage, death during postdiapause development, incomplete ecdysis, malformation in emerged adults, to delayed emergence. The impact of the cold exposure was cumulative and there was no 'repair' of the cold injuries at 0 °C and, most importantly, the observed cold injuries were not caused by freezing. The study on *M. configurata* shows clearly that the effects of cold exposure can be 'delayed' and can cause death in subsequent developmental stages. Based on the results with *M. configurata* Turnock *et al.* (1983) suggest that there are temperature zones, which produce different effects on

Table 4.5. *Mean supercooling points ($\pm$ standard error) and*
LT_{50} (°C) with 95% fiducial limits of the peach-potato aphid
Myzus persicae

Age group	Mean SCP $\pm$ SE	LT_{50} (with 95% FL)
First instar	-26.6 ± 0.2	-8.1 (-7.2 to -9.0)
Adult	-25.0 ± 0.1	-6.9 (-6.0 to -7.7)

FL, fiducial limits; SCP, supercooling point; SE, standard error.

invertebrates. Survival and development are affected in the cold injury zone, and the effects are cumulative. There is no repair of cold injury within this zone, the lower boundary of which is the supercooling point. There is no cold injury in the neutral zone (within the limits of exposure periods that might be experienced in the natural environment) but there is no repair of cold injuries already incurred, and previous levels of cold-hardiness are maintained. Finally, there is an active zone in which no cold injury occurs, minor cold injuries can be repaired and changes in cold-hardiness may occur.

Mortality in *Cryptopygus* and *Mamestra* occurs over winter in field populations and in laboratory samples maintained at subzero temperatures above their supercooling point for long periods of time, but survival of short exposures at temperatures close to the supercooling point is very high; for instance over 80 per cent in some winter samples of *Cryptopygus* survived after 24 hours at -15 °C (Block 1987). It has recently been shown that, although aphids are capable of extensive supercooling, this attribute affords little or no protection against low temperatures and even within the time limits of a supercooling experiment at 1 °C/minute, some aphid populations show very high mortality above their supercooling point (Table 4.5; Knight *et al.* 1986b; Bale *et al.* 1988). Increasing the period of exposure at a particular temperature may increase the level of prefreeze mortality for these insects rather than increasing prefreeze relative to freezing deaths.

The influence of time on the likelihood of death at low temperature is in some ways the most important aspect of insect overwintering, but it is also one of the least researched. It is self-evident that most freezing intolerant insects live in environments where the supercooling ability of a proportion of the population exceeds the minimum winter temperatures, or where sufficient individuals seek out effective hibernaculae such that they are able to avoid periods of extreme cold; if this was not the case the insects would have become extinct or limited in their distribution long ago. From this it can be deduced that the greatest threat of low temperature to insects in nature is the length of time that they have to spend in the supercooled state. While it is certainly an understatement to say that 'more work is needed on this subject' it must also be recognised that, experimentally, the 'time factor' constitutes a major challenge to investigators. Overwintering insects are subject to fluctuating subzero temperatures and variable cooling rates. Most 'time' experiments are conducted at constant temperatures and virtually all supercooling experiments utilise a cooling rate of about 1 °C/minute. Clearly it is impossible to represent field conditions in the laboratory except for short periods of time, but it remains a primary task for insect ecologists and cryobiologists to investigate the effect of the length of cold exposure on low temperature mortality and to identify the causes of death under such conditions.

Environmental triggers for cold-hardening

Most of the strategies adopted by insects to overcome the problems of living in an inconstant environment (diapause, polymorphism, rhythmic behaviour, migration, cold-hardiness) involve physiological, biochemical or behavioural responses to environmental stimuli, often described as 'cues' or 'triggers'. In the majority of these survival strategies, photoperiod is regarded as the most reliable 'indicator' of changing season. In a particular locality the photoperiod on a given day will not vary from year to year, whereas temperature varies in an unpredictable manner. It can be argued (Baust 1982) that as a trigger for cold-hardening, temperature provides a more reliable cue due to its indicative rather than predictive nature, i.e. temperature indicates the local conditions at that moment in time, whereas decreasing photoperiod predicts that summer is changing to autumn and will be followed by winter.

Evidence is available for the primary importance of temperature as a trigger for cryoprotectant synthesis in both freeze tolerant and intolerant insects, but much less is known about the production of ice nucleating agents and antifreeze proteins. In some species the environmental cue is more specialised, such as photoperiod (Duman 1980; Horwath and Duman 1982), moisture (Rojas *et al.* 1986) and dietary carbon source (Baust and Edwards 1979), and in these cases is usually involved in the synthesis of an individual cryoprotectant or protein.

The small carabid beetle *Pterostichus brevicornis* overwinters in decaying tree stumps and was found to accumulate only one polyol cryoprotectant, glycerol, in quantities sufficient to increase winter cold-hardiness

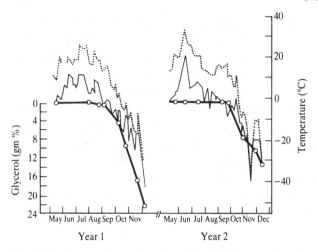

Fig. 4.15. Seasonal variations in daily high ($\cdots\cdots$) and low (————) temperatures and haemolymph glycerol levels (O—O) of *Pterostichus brevicornis* (after Baust 1982).

(Baust and Miller 1970, 1972). The glycerol content of adult beetles in relation to seasonal variations in daily high–low temperatures over 2 years is shown in Figure 4.15 (Baust 1982). Glycerol content increased immediately after the first frost in both years, although the onset of winter occurred 1 month apart, on 23 August in year 1 and 25 September in year 2. Baust has argued that photoperiod would be an unreliable cue during the autumn in Alaska because the extended periods of twilight provide an effective 24 hours of 'daylight' whereas, over the same 2–3 months, first frost may occur at any time, such that a greater survival benefit is obtained from reliance on a prestress temperature trigger.

Field populations of *P. brevicornis* collected before the first frost did not accumulate any glycerol when maintained in the laboratory for up to 4 weeks at 5 °C. However, glycerol was detected after only 2 days at 0 °C, with higher levels after the same period of exposure to −5 and −10 °C and with increasing periods of exposure. The results were similar in light:dark periods of 0:24, 12:12 and 24:0 (Baust 1982). It is also interesting to note that continued monitoring of *P. brevicornis* showed that glycerol content varied directly with temperature changes throughout winter but with no loss of cold-hardiness. Peaks of glycerol in December and March reflected the peak low temperatures at the same time. However, a rapid temperature decrease in late December of over 30 °C in 5 days was followed by a 'lag phase', in which glycerol content did not increase until mid-January. Baust observed that there was no snow cover on the tree stumps in early December and snow that fell in mid-December was largely melted by March, so that the beetles were exposed at these times to variations in ambient temperature. In late December there was a snow cover of 0.5 m, which: 'damped the velocity of thermal fluctuations in the stump' so the low temperatures occurring at the same time did not cool the insulated microhabitat of the beetles until about 2 weeks later, hence the time lag in the increase of glycerol in the beetles (Baust and Miller 1970). Laboratory experiments also revealed a positive 'cryotaxic' response in the beetles, which showed a mean temperature preference of $-5.5\,°C$ in midwinter (with some individuals showing a preference of $-11\,°C$) and $13.3\,°C$ in summer, suggesting that in winter beetles seek a colder and more temperature stable microclimate, which would be advantageous in a species such as *P. brevicornis*, which loses glycerol after temporary exposure to fluctuating high temperatures.

There are many reports of increasing cold-hardiness under the influence of decreasing temperature and some of these have included experiments to identify the optimal temperature for acclimation. Some reviews have concluded that acclimation is more effective at temperatures just above zero $(3-5\,°C)$ rather than at higher or subzero temperatures (Ring 1980). There is some evidence that the temperatures required to induce the acclimation response may reflect the regional climate, alpine and polar insects having lower thresholds than temperate species. For instance, the beech weevil, *Rhynchaenus fagi*, collected in October showed an increase in supercooling of more than $4\,°C$ $(-20.8$ to $-25.3\,°C)$ when maintained at $5\,°C$ for 2 weeks (Bale and Smith 1981), equivalent to the most cold-hardy individuals collected from the field in January (Bale 1980). Supercooling of overwintering temperate aphids is unaffected by acclimation (Knight and Bale 1986) but these insects show extensive prefreeze mortality with the acclimation response manifest as a depression of the LT_{50}. Aphids become more cold-hardy when maintained at 10 and $5\,°C$ (compared to $20\,°C$) but die at $0\,°C$ (Table 4.6; Bale *et al.* 1988). In contrast, the optimal temperature for cryoprotectant synthesis appears to be below $0\,°C$ for a number of freezing intolerant Collembola and mites from cold climates reviewed by Sømme (1982). Thus, the mean supercooling point of the alpine collembolan *Tetracanthella wahlgreni* decreases from -6.8 to $-31.6\,°C$ from summer to winter with glycerol accumulating at -5 and $-10\,°C$ but not at $0\,°C$. Two species of cryptostigmatid mite from the same habitat also showed a greater increase in glycerol content at $-5\,°C$ than at $0\,°C$ (Sømme and Conradi-Larsen 1977). Similar results have been obtained with the Antarctic mite, *Alaskozetes antarcticus*, where

Table 4.6. *LT$_{50}$ (°C) and 95% fiducial limits of the peach-potato aphid*
Myzus persicae *reared at 20, 10 and 5 °C*

	Rearing temperature (°C)		
Age group	20	10	5
First instar	−8.1	−11.8	−16.4
	−7.1 to −9.2	−11.0 to −12.5	−14.5 to −19.4
Adult	−6.9	−11.1	−11.6
	−6.0 to −7.7	−7.0 to −14.7	−7.5 to −13.7

glycerol production was enhanced at −5 and −10 °C compared to 0 °C
(Young and Block 1980), while −10 °C was the most effective temperature
in increasing supercooling in *Cryptopygus antarcticus* (Cannon 1986c).

In species with a multifactorial cryoprotectant system it is now apparent
that the synthesis of individual compounds may required different cues.
Overwintering freeze tolerant larvae of the goldenrod gall fly, *Eurosta
solidaginis*, accumulate four cryoprotectants: glycerol, sorbitol, trehalose
and fructose, with the polyols occurring in the highest concentrations
(Baust and Lee 1981). Sorbitol synthesis has a temperature trigger, with
an upper threshold at 10 °C and maximum accumulation at 0 °C (Rojas
et al. 1983). However, glycerol content increases in natural populations
before the first frosts and it is clear that its production is not dependent
upon chilling and that temperatures below 10 °C may in fact inhibit
glycerol synthesis (Baust and Lee 1981; Rojas *et al.* 1983). Collection of
plant galls from autumn to winter revealed a strong correlation between
the gall water content, which decreased from 65 to 20 per cent between
November and January, and the glycerol content of the larvae, which
increased from 10 to 21 μg/mg over the same period (Fig. 4.16; Rojas
et al. 1986). Larval water content remained constant throughout the
autumn/winter season at around 60–65 per cent but larval weight
decreased by nearly 50 per cent (65–37 mg) and was correlated with the
decreasing gall water content. The interesting hypothesis that emerged
was that the progressive desiccation of the gall may serve as an indirect
trigger for glycerol synthesis. Further evidence was gained by the collection in October of 'dry' galls that had begun their 'winter' senescence
early, and a sample of normal green galls. Larvae taken from the green
(wet) galls contained 4.6 μg/mg of glycerol whereas dry gall larvae were
significantly higher, at 7.5 μg/mg. At present, the relationship between the
gall water content and glycerol accumulation in the larvae remains a

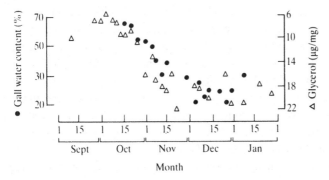

Fig. 4.16. Seasonal variations in *Eurosta solidaginis* (third instar larvae) glycerol and water ($\triangle$) contents and goldenrod (*Solidago* sp.) gall water content ($\bullet$) (after Rojas *et al.* 1986).

strong correlation; laboratory experiments with larvae exposed to various humidities did not produce conclusive evidence to support the hypothesis and it was concluded that other factors, such as diet and the stage of larval development may also be involved in the glycerol trigger (Rojas *et al.* 1986). A previous study (Young and Block 1980) indicated that although low temperature was the cue for glycerol synthesis in the mite *Alaskozetes antarcticus*, a higher level of glycerol was produced at low relative humidities. The desiccation factor may be important in *Eurosta* and the equivocal nature of the laboratory data may simply reflect a problem encountered with many aspects of cold-hardiness research, namely the creation in the laboratory of environmental conditions sufficiently similar to the natural habitat to induce the physiological and biochemical changes observed in field-collected specimens.

In terms of their discovery by insect cryobiologists, ice nucleating agents and antifreeze proteins have a more recent history than cryoprotectant and antifreeze polyols and sugars. Much of the research on these two groups of haemolymph proteins has concerned their structure and function and as yet few species have been studied from the viewpoint of environmental triggers for their synthesis.

Ice nucleating agents were first described by Zachariassen and Hammel

(1976) in the beetle *Eleodes blanchardi* and have since been recorded in a number of species including the hornet, *Vespula maculata* (Duman and Patterson 1978) (see p. 83). A similar pattern of results has been obtained from various experiments on the two species. Both are freezing intolerant in summer, the supercooling point being slightly lower in summer than winter in *E. blanchardi* ($-9.3\,°C$ compared to $-6.3\,°C$) and about the same in *V. maculata* (-3.8 and $-4.6\,°C$). Both species become freeze tolerant in winter, with the lower lethal temperature below the super-cooling point. The supercooling points of NaCl and glycerol were raised, in comparison with solutions containing summer haemolymph, by the addition of winter haemolymph from freeze tolerant *E. blanchardi* (Zachariassen 1980; Zachariassen and Hammel 1976). Furthermore, 100-fold dilution of winter haemolymph from *V. maculata* reduced its supercooling point by only $0.1\,°C$ compared to an undiluted sample. Clearly the ice nucleators are highly effective.

It seems that nearly all freeze tolerant insects contain ice nucleating agents (Duman and Horwath 1983) and that the species are usually freezing tolerant only in winter. The strong correlation between freeze tolerance and nucleator activity in winter suggests that decreasing temperature in autumn is the most likely trigger for ice nucleator synthesis.

However, it is worth stressing, that, in most species, the winter activity of nucleators is part of a multifactorial process of cold-hardening. For instance, the summer glycerol content of *V. maculata* of 12 mg per cent increases to 3788 mg per cent in winter (Duman and Patterson 1978) and the winter freeze tolerance of *E. blanchardi* is lost after 2 weeks at $20\,°C$ even though the supercooling point does not change, implying that some other essential component of the freeze tolerant condition has been lost (Zachariassen and Hammel 1976).

Antifreeze proteins are able to depress the haemolymph and whole-body supercooling points of insects and may be of greater importance to freezing intolerant insects than polyols and sugars. In some species, such as *Dendroides canadensis* (Coleoptera: Pyrochroidae), which has been observed to change its strategy from freeze tolerant to intolerant in different years (Horwath and Duman 1984), antifreeze proteins act with glycerol and sorbitol to depress the supercooling point of the haemolymph and whole insect (the latter effect depending on the presence or absence of the ice nucleator). In other species, such as the overwintering freezing intolerant larvae of *Meracantha contracta*, the antifreeze effect appears to be based entirely on proteins – no polyols or sugars were detected (Duman 1977a,b).

The level of antifreeze protein in the haemolymph of these species increases in early autumn and decreases in spring, suggesting that environmental cues indicative of seasonal changes control the production and loss of the proteins (Duman and Horwath 1983). Acclimation experiments have shown that low temperature and/or short photoperiods induce antifreeze synthesis in *M. contracta* (Duman 1977c), *D. canadensis* (Duman 1980; Horwath and Duman 1982) and *Tenebrio molitor* (Patterson and Duman 1978).

Although the information on antifreeze proteins is as yet based on a small number of species, there is evidence that their production arises primarily from a predictive (photoperiod) rather than an indicative (temperature) cue and Duman (1982) has argued that the value of a system utilizing antifreeze proteins would be enhanced if synthesis occurred in anticipation of low winter temperatures, so preventing injuries from sudden early frosts. Studies on *M. contracta* and *D. canadensis* support this view. *M. contracta* larvae overwinter in decomposing logs that, together with snow cover, provide some insulation from the prevailing climate. However, this microhabitat can become quite wet in winter and, in the absence of other cryoprotectants, the depression of the haemolymph freezing point to around −5 °C, which lessens the possibility of inoculative freezing across the cuticle, appears crucial to survival (Duman 1977a,b).

The situation with *D. canadensis* is more complex and is discussed fully in the next section (comparison of cold-hardiness strategies). Again the larvae overwinter under the bark of dead trees. In some years the larvae are freeze tolerant for most of the winter (although they may become temporarily freezing intolerant during midwinter warm periods), while in other years the larvae are freezing intolerant throughout winter, attributable to the presence or absence of ice nucleators. However, antifreeze protein synthesis (and that of polyols) occurs every year irrespective of the freeze tolerant or intolerant nature of the overwintering larvae (Duman 1982; Horwath and Duman 1984). It seems likely that photoperiod is an important component in triggering the production of antifreeze proteins in *D. canadensis* (and low temperature may regulate the level of production) whereas polyol (glycerol and sorbitol) synthesis is 'switched on' each winter by low temperature, and ice nucleator content depends on the severity of cold experienced by larvae in the previous winter. It has been shown that the critical temperature for the induction of antifreeze proteins in *Dendroides* is between 10 and 15 °C (a relatively high temperature threshold) while the critical photoperiod lies between a light:dark cycle of 10:14 and 11:13 at 20 °C (Duman and Horwath 1983).

Additionally, the protein synthesis occurs with a circadian frequency (Horwath and Duman 1982).

Ecologically it may be advantageous for a freezing intolerant insect to possess two antifreeze systems, one of which (antifreeze proteins) provides initially early-season protection and is triggered by a predictive cue such as photoperiod, and the second, polyols and sugars, extends supercooling capacity in relation to an indicative temperature cue and, in combination, produces a stable cold-hardy supercooled state. This scenario would apply to the years in which *D. canadensis* overwinters as freezing intolerant larvae. The photoperiodic induction of antifreeze proteins in autumn is also clearly advantageous to *D. canadensis* in those years when it overwinters as freeze tolerant larvae, protecting the insects from freezing injury in the autumn to early winter period before they achieve their 'predetermined' freezing tolerant state, in combination with the low temperature-induced polyols and sugar cryoprotectants; the same protection would continue through midwinter thaws and into the spring, when the freeze tolerant condition is lost (Duman 1980; Duman *et al.* 1982; Horwath and Duman 1984).

The schematic presentation of the principal adaptive strategies of overwintering insects (see Fig. 4.1; Baust 1982) indicates a primary role for cryoprotectant and antifreeze polyols and sugars, antifreeze proteins and ice nucleating agents. Furthermore it is now abundantly clear that Baust's conclusion that: 'it is unlikely that any single environmental cue (trigger) can provide the definitive signal to initiate each of the multifactorial strategies demonstrated during (cold) hardening' is undoubtedly true. In the light of the information in this section it would be surprising if winter cold-hardiness was based on a single cue.

What emerges from this apparent milieu of environmental stimuli and biochemical responses is the development of highly organised alternative overwintering strategies that are multifaceted both in terms of their components and the factors governing the initiation and regulation of their synthesis. Many freezing intolerant insects contain antifreeze proteins and polyols and sugars, and these are also found in freeze tolerant species, with the addition of ice nucleating agents. The combined effect in both strategies is to provide an adequate level of cold-hardiness in autumn and early winter, based mainly on antifreeze proteins; maximum cold-hardiness in mid- to late winter, when polyol and sugar antifreezes/cryoprotectants and ice nucleating agents play a decisive role; and continued protection until spring from the antifreeze proteins, which remain active after the polyols and ice nucleators have been lost. Viewed

Table 4.7. *Haemolymph melting point, hysteresis freezing point, whole-body supercooling point and lower lethal temperature (all °C) of larvae of* Dendroides canadensis *in different winters (after Horwath and Duman 1984)*

Winter	MP	HFP	SCP	LLT
1977–8 (Feb)	-4.68 ± 0.79	-7.54 ± 0.7	-11.2 ± 1.0	-28.0
1978–9 (Feb)	-5.46 ± 1.03	-11.06 ± 2.68	-12.0 ± 1.3	-25.5
1981–2 (Feb)	-2.54 ± 0.18	-5.70 ± 1.2	-27.3 ± 0.7	-28.0

MP, melting point; HFP, hysteresis freezing point; SCP, supercooling point; LLT, lower lethal temperature.

against this background it is clearly beneficial for insects to receive both predictive and indicative signals from their environment, to trigger the different processes that occur in sequence to promote and maintain winter cold-hardiness.

Comparisons of cold-hardiness strategies

Historically, tolerance and intolerance of freezing have been regarded as distinct strategies with different physiological mechanisms. The titles and content of research papers, reviews and conference sessions have followed this trend and as such have emphasised the differences between tolerant and intolerant species while overlooking the many similarities between the two strategies of overwintering. In fact, the differences (apart from the action of ice nucleating agents) lie not in the biochemical components of tolerant and intolerant insects, which are essentially the same, but in the function of the compounds in the two modes of overwintering.

Species that change strategy

It might be expected that, within a single species, tolerance and intolerance are mutually exclusive strategies, but at least two species of beetle which overwinter as larvae, *Dendroides canadensis* and *Cucujus clavipes* (Coleoptera: Cucujidae) are known to be *either* freezing tolerant *or* freezing intolerant in different winters, but do not, apparently, show a mixed response within the same winter (Horwath and Duman 1984). Larvae of *D. canadensis* were studied over five winters, 1977–82, in northern Indiana. The first two winters 1977–8 and 1978–9, were very cold, the next two were more mild, while the winter of 1981–2 was as severe as those in 1978 and 1979. The results from the three coldest winters are summarised in Table 4.7. In winters 1977–8 and 1978–9 the larvae were freeze tolerant, containing ice nucleating agents with whole body supercooling points between -8.0 and $-12\,°C$; there was also a clear protein antifreeze thermal hysteric effect. The lower lethal temperature of the larvae was $-28\,°C$. By contrast, in winter 1981–2 the larvae were freezing intolerant, with whole body supercooling points in the region of $-26\,°C$ corresponding to the lower lethal temperature. Interestingly, the larvae still showed a thermal hysteresis effect and the only difference was the activity of ice nucleating agents, which functioned in the freeze tolerant larvae to limit supercooling to a few degrees below the 'hysteric freezing point'. *D. canadensis* takes two or more years to complete its life cycle, so each larva will overwinter more than once; as only the later instars were used in the study, all the experimental larvae had experienced at least one previous winter. Measurements of haemolymph supercooling points from larvae collected in the two mild winters indicated that the change of strategy occurred during the 2-year period and that the conditions experienced by the early instars in their first winter may influence the mode of overwintering adopted in the second winter. For instance, in the first mild winter, 1979–80, the haemolymph supercooling points in October and February were -7.1 and $-10.8\,°C$, similar to the supercooling points in the previous two cold years when the larvae were freezing tolerant. In the second mild winter, 1980–1, (when the experimental larvae would have experienced the previous mild winter as early instars) the October and February haemolymph supercooling points were -18.2 and $-25.4\,°C$, in this case similar to the supercooling points of larvae examined in 1981–2 when the larvae had become freezing intolerant. Long term monitoring will reveal if the change of strategy in *D. canadensis* is permanent or if there is some switch mechanism that determines the freeze tolerant or intolerant nature of mature 'second winter' larvae depending on the 'severity of cold' experienced by the early instars a year earlier.

The ability of a species to change strategy was an unexpected discovery but it adds weight to the idea that the key biochemical components, such as polyols and antifreeze proteins, are common to most overwintering insects but that their roles depend on the presence or absence of ice nucleating agents, which determine the freeze tolerant or intolerant nature of the insect.

Designation of species as freeze tolerant or intolerant

While the control mechanisms for the switch of strategy in *D. canadensis* are unknown, it is interesting to note that the supercooling points of the larvae were determined at the same rate (0.2 °C/minute) throughout the study, and there can be no doubt, therefore, that the specimens showed a different physiological response in different years. Recently, Baust and Rojas (1985) have considered some of the assumptions and inaccuracies that may arise in the designation of species as freeze tolerant or intolerant and the assessment of survival, when insects are cooled at an arbitrary and uniform rate, usually 1 °C/minute. The adoption of a standard cooling rate allows comparisons between different species and investigators on the grounds that different cooling rates may affect the observed super-cooling point. Baust and Rojas argue that, while the supercooling point may change, more importantly, cooling rate may influence the numbers of insects that survive freezing, and therefore the designation of the species as tolerant or intolerant.

Measurement of supercooling points is a non-invasive technique that detects the heat liberated (latent heat of fusion) as the body fluids undergo a phase change from the liquid to solid state. Baust and Rojas stress that the onset of the freezing exotherm provides a clear indication of when the freezing process is initiated, but that it does not represent the total freezing of all available body water, even after the rebound of the cooling curve to the initial freezing temperature.

The supercooling point has been widely used for two purposes: (i) as the temperature at which the ability of an insect to survive freezing can be assessed (and then classified as freeze tolerant or intolerant); and (ii) as the temperature that represents the lower lethal limit of a freezing intolerant insect. The views expressed by Baust and Rojas suggest that neither of these interpretations may be correct if the conclusions are derived from experiments conducted at a single and arbitrary rate of cooling. Information available from extensive research on the cryo-preservation of cells and tissues of medical and veterinary importance indicates that the cooling rate (which in this context can be regarded as synonymous with freezing rate) and warming rate (synonymous with thawing rate) of the storage protocol are crucially important in determin-ing the survival and viability of the products after freezing. It seems likely that the ability of an insect to survive freezing will depend on a series of 'optimal conditions', of which the cooling and warming rate are two of the most important. It is equally likely that the 'optimal rate' will differ between species and may be faster or slower than the usual standard of 1 °C/minute. In one of only a few studies that has attempted to determine the optimal cooling rate for species, a value of 0.1 °C/minute was found for the freeze tolerant larvae of the Antarctic midge, *Belgica antarctica*

(Diptera: Chironomidae) (Baust 1980). It is probable that cooling rate will affect the supercooling point and survival after freezing independently and previous studies that have shown cooling rate to have only a marginal influence on the observed supercooling point (Salt 1966a) may not have included the required cooling rates to reveal any differences in survival below the supercooling point. The cooling rate of 1 °C/minute for species currently regarded as freezing intolerant may be suboptimal such that at their optimal cooling rate they may show some tolerance of freezing. Conversely, the cooling rate of 1 °C/minute may be coincidentally the optimal rate for survival for some species that are then classified as freeze tolerant, whereas at other rates the same species may not survive the freezing process. Some evidence for this view has been gained from recent cooling and warming rate experiments with the freeze tolerant larvae of the gall fly, *Eurosta solidaginis*. When larvae were cooled at 0.1, 1, 5 and 10 °C/minute to −40 °C (with supercooling points between −8 and −10 °C), and then warmed at 1 °C/minute to +20 °C, survival to adult emergence was 77, 37, 13 and 7 per cent respectively, compared to 67 per cent in a control group; interestingly, 1 °C/minute was not the optimal cooling rate for survival. When larvae were cooled at 0.1, 1, 5 and 10 °C/minute to −40 °C and then warmed at 0.1 or 10 °C/minute, survival was again highest at the slowest cooling rate but, at each cooling rate, survival was higher at the slower warming rate (Bale *et al.* 1989b).

Although at first sight it may appear that tolerance or intolerance of freezing will vary according primarily to the cooling rate, survival of freezing is dependent upon the interaction between the rate of cooling to the supercooling point and the rate of warming to the thawing temperature, and both rates will have their own optimum. The combination of an optimal cooling rate and a suboptimal warming rate, and vice versa, will not reveal the true ability of a species to survive freezing. The preceding discussion has described cooling and warming regimes as rates that may imply a constant decrease or increase in temperature per unit time. In fact, recent experiments with various developmental stages of the flesh fly, *Sarcophaga crassipalpis* (Diptera: Sarcophagidae), have shown a high 'cold shock' mortality when pharate adults of non-diapausing flies are exposed directly to −10 °C for 2 hours in insects that supercool to −23 °C, and a marked increase in survival to adult emergence if the pupae are acclimated at 0 °C for 2 hours before exposure to −10 °C (Chen *et al.* 1987). Gradual cooling from 25 to −10 °C also produced a very high mortality, and survival at −10 °C was found to be directly related to the absolute time of prior exposure to 0 °C. Subsequent experiments showed the optimum temperature for this short term acclimation to be between 0 and 6 °C. Chilling at 0 °C also increased survival at − 10 °C in other developmental stages beyond the third instar in both diapausing and non-diapausing strains, with the exception of the phanerocephalic pupal stage typical of diapausing pupae, which can tolerate − 10 °C without pretreatment at 0 °C (Lee and Denlinger 1985). Interestingly, the 2-hour cold treatment at 0 °C increased the glycerol concentration by a factor of two to three compared to unchilled controls and was sufficient to account for the increased osmolality of the haemolymph for pretreated flies. This study adds another category of low temperature mortality (cold shock) to those of: (i) the lower lethal limit of freezing tolerant species; (ii) freezing intolerance; and (iii) groups such as aphids, which die above their supercooling point during or after cooling at 1 °C/minute. Apart from suggesting an additional cryoprotective role for glycerol, which, in *S. crassipalpis*, can be produced very rapidly, the results also indicate that both the level of mortality and the cause of death can be affected by the rate or regime of cooling that is applied to experimental insects.

Investigations on the effect of cooling and warming rates on low temperature mortality are beset by the same type of underlying problem that affects studies on the long term cold exposure – while it is desirable to compare mortality at a range of cooling rates and to expose insects to fluctuating as well as constant subzero temperatures in 'time' experiments under laboratory conditions, it is only the cooling and warming rates and the temperature and periods of exposure experienced in the overwintering microhabitat of the insect that will determine its freeze tolerance, intolerance or other response to low temperature. In the seasonal sequence from autumn through winter to spring, there are regular diurnal and longer term patterns of temperature change. While the advantages of adopting a standard cooling rate are accepted, it must also be recognised that many of the general principles of insect cold-hardiness are based on a rate of cooling that is rarely, if ever, experienced by natural populations in their overwintering sites.

What is survival?

The classification of a species as either freeze tolerant or intolerant is based on its ability to survive freezing. In most experiments this assessment is made at a fixed time interval after cooling the specimen at 1 °C/minute, and allowing it to thaw gradually, and at an uncontrolled rate, to ambient. Even if the cooling and warming rates are optimal, the selection of an appropriate time interval after exposure and related criteria for determining survival are crucial. For instance, survival estimates are often made only a few hours after warming and observations do not usually extend beyond one week; ability to walk, feed or respond to a simple stimulus are commonly used tests for survival (Baust and Rojas 1985). Two aspects of assessing low temperature survival require careful consideration. First, how long after exposure should assessments be made, and secondly, can an organism be regarded as a survivor if, as a result of freezing (or in some cases, only cooling) it is unable to reproduce?

The carabid beetle, *P. brevicornis*, is freeze tolerant as an adult to around −80 °C (Miller 1969). Short term survival is high but after a week all the experimental beetles are dead with no mortality in the controls. Aphids are neither freezing tolerant nor intolerant but show extensive prefreeze mortality. After cooling a large sample of aphids to a temperature above their supercooling point there are three distinct categories of response observed 1 hour after exposure: dead (usually few), injured, and alive (presumed healthy). Those aphids that are clearly injured usually die within 24 hours, with a few individuals surviving for 48 or 72 hours; for this reason 'final' mortality counts are made 3 days after the experiment (Knight *et al.* 1986b). However, longer term monitoring of first instar nymphs of the nettle aphid, *Microlophium carnosum* (Hemiptera: Aphididae), which were healthy and resumed feeding after exposure to −20 °C, showed that no individual survived to the adult stage (Pullin and Bale 1988). Mortality increased with time in both these studies and insects that were apparently unharmed by their cold exposure and responded normally to various tests and would have been classified as survivors at that moment in time, died only a few days later. In fact, a recent study on the low temperature survival of adult *Myzus persicae* (McLeod 1987) indicated that exposure to −15 °C had little effect on mortality, the LT_{50} (time) being 68 hours. This result is in marked contrast to other experiments on *M. persicae* (Bale *et al.* 1988) that showed the LT_{50} (temperature) of acclimated and unacclimated adult aphids after cooling at 1 °C/minute to be −11.1 and −8.1 °C, respectively, when examined 3 days after exposure. Whereas McLeod assessed mortality after 45 minutes at room temperature and only those aphids that showed no movement were classified as dead, Bale *et al.* did not make their first mortality assessment until 24 hours after exposure, for the very reason that most aphids were not killed immediately by their cold exposure but died some time later after their return to the higher temperatures required for normal metabolic activity. A more extreme example of this type of effect is shown by larvae of *Eurosta solidaginis* cooled to −40 °C at different rates (0.1, 1, 5 and 10 °C/minute); at 10 °C/minute, although all the larvae were alive 48 hours after cooling, only 7 per cent survived to adult (Bale *et al.* 1989b).

In some cases, the deleterious effects of freezing or subzero prefreeze exposure may be much delayed so that the insect dies at a later stage through some indirect effect of cold. For instance, pupae of *Mamestra configurata* exposed for long periods of time to subzero temperatures above the supercooling point, either died in diapause, died during postdiapause development, failed to moult, emerged as malformed adults or emerged late (Turnock *et al.* 1983).

Most insects overwinter only once in their life. Overwintering eggs, larvae or pupae usually develop to adult in the following spring or summer and then reproduce. Species that overwinter as adults are often in a state of reproductive diapause that terminates in mid- to late winter with reproduction in spring. Reproduction can therefore be regarded as the only true test of survival; what contribution does an individual make to its species if it survives through winter but is rendered sterile by its cold exposure? The problems associated with this interpretation of survival have been recognised by Baust and Rojas (1985). It is more convenient to complete a survival assessment within a few days of exposure based on the proportions of the sample that are living or dead. If the test specimens are to be maintained for subsequent observations on reproductive capability and fecundity, the laboratory may require extensive rearing facilities and the investigator will have to be familiar with the biology of the species, particularly the photoperiodic and temperature requirements for diapause and postdiapause development. It is evident that few studies have included any assessment of reproduction in estimates of survival and the value of this information must be viewed in this context.

Ecophysiological approaches to insect cold-hardiness

The study of insect cold-hardiness combines areas of physiology, biochemistry and ecology and is concerned with all those events and processes governed by low temperature that influence and ultimately determine survival or mortality in the natural environment.

The introduction to this book has stressed the basic dichotomy of research on insects and low temperature between laboratory-based techniques in physiology and biochemistry and field-based population ecology. With reference to the dominant areas of cold-hardiness research, Storey (1984) observed that physiological ecologists have examined hundreds of species, determining their freezing tolerance or intolerance by measuring supercooling, freezing and melting points of whole organisms and body fluids, and the type and amount of cryoprotectant, while biochemists and biophysicists have concentrated on problems such as the structure of water in cells, cryoenzymology and the protein chemistry of antifreeze molecules. What becomes immediately apparent from Storey's accurate appraisal is the absence of comparative studies that relate laboratory estimates of cold-hardiness, such as the supercooling point or LT_{50}, to temperatures experienced in overwintering sites, and to some measure of the level of mortality or survival in winter in the natural environment. Unless this is done, it is impossible to assess the efficacy (or even the requirements) of the envisaged cold-hardiness strategy in nature, or to make the transition from hypothesis to fact. At the other end of this spectrum of research, many studies by population ecologists have identified winter cold stress as an important mortality factor of insect populations, but rarely has a research programme determined the freeze tolerant or intolerant status of the species, or considered the value of indices of cold-hardiness in predictive models.

The cold-hardiness of aphids has been a subject of considerable interest in recent years because many applied entomologists have observed an apparent relationship between the severity of winters in successive years and the abundance of aphids in the following summers; mild winters favoured survival but mortality was high in winters with severe frosts. Much of the early research on aphid cold-hardiness was based largely on supercooling point determinations. The results were interesting because they showed that aphids were atypical in their supercooling characteristics. Most freezing intolerant species, such as the beech leaf mining weevil, *Rhynchaenus fagi*: (i) acclimatise for winter by depressing their supercooling point; (ii) accumulate antifreeze agents in response to low temperature; (iii) freeze above their supercooling point when wet, by inoculation of freezing through the cuticle; and (iv) freeze above the inherent supercooling point when feeding, through the action of gut (food) nucleators (Bale 1980; Bale and Smith 1981). By contrast, aphids did not show any winter increase in supercooling; the supercooling point was consistently below $-20\,°C$ throughout the year. Nor did they show any supercooling acclimation response when maintained at low temperature. Additionally, only a small proportion of aphids froze above their supercooling point when inoculated with surface moisture and feeding (on sap) did not reduce supercooling (O'Doherty and Bale 1985; Knight and Bale 1986).

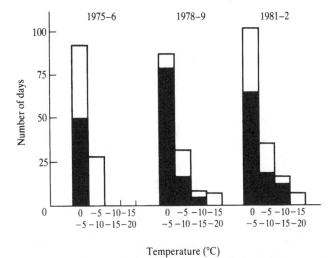

Fig. 4.17. Variation in the number and severity of frost days in mild (1975–6) and severe (1978–9, 1981–2) winters in northern England. ■ Air screen minimum, □ grass minimum (after Bale *et al.* 1988).

On the basis of their supercooling ability, aphids can be regarded as cold-hardy freezing intolerant insects, with winter supercooling points usually between −20 and −25 °C. Examination of temperature records for northern England over nine successive winters from 1975 revealed that in mild winters the temperature did not fall below −10 °C (even at the soil surface) and that in the coldest winters temperatures below −15 °C were rare; on no occasion did the temperature fall to −20 °C (Fig. 4.17) (Bale *et al.* 1988). By comparing supercooling ability to environmental temperatures it might be expected that aphid survival would be relatively high in most British winters. In fact, detailed population monitoring of the peach-potato aphid, *Myzus persicae*, and the grain aphid, *Sitobion avenae* (Hemiptera: Aphididae), has shown winter population declines consistently above 97 per cent in both species with abrupt increases in mortality following sudden decreases in temperature (Fig. 4.18; Harrington and Cheng 1984; Knight *et al.* 1986b). The field results suggested that aphids were dying at temperatures much above their supercooling points from the effects of a cold-related prefreeze mortality factor. Subsequent laboratory experiments have confirmed the importance of prefreeze mortality in aphids (Bale *et al.* 1988).

The aphid research highlighted inconsistencies between: (i) A laboratory determined index of cold-hardiness such as the supercooling point;

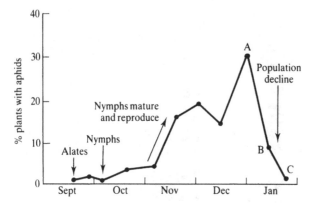

Fig. 4.18. Population sequences of *Sitobion avenae* on winter barley in winter 1984–5. Aphid density declined from 30 to 1 per cent of plants infected (1 m to 30 000/ha) from A to C when the lowest grass minimum was −8.1 °C between A and B and −9.7 °C between B and C. The mean supercooling point was < −20 °C (after Knight *et al.* 1986b).

(ii) environmental and microclimate temperatures experienced by over-wintering insects; and (iii) winter mortality assessed in the field. It is clearly beneficial to adopt an integrated approach to the study of insect cold-hardiness and in line with this theme, Bale (1987) has recommended a protocol for research in five stages:

1. Make a preliminary hypothesis on the freezing status (tolerant or intolerant) of the species and assess the lower lethal temperature or supercooling point as appropriate at a standard rate of cooling.
2. Record annual, seasonal and microclimate temperatures.
3. Monitor population density of the subject species in winter in either natural or, if more convenient, caged populations.
4. Assess mortality at low temperature at prefreeze, freezing (supercooling point) and postfreeze temperatures, varying the cooling and warming rates, periods of exposure and phases of acclimation.
5. Investigate the biochemical and physiological causes of death and adaptations for survival.

If, for instance, an insect shows extensive supercooling (1) that far exceeds the temperatures it is known to experience in its overwintering site (2), but dies in large numbers in winter (3), then further studies in the laboratory (4) can quantify the mortality above and at the supercooling point in relation to the cooling rate, minimum temperatures experienced and the periods of exposure. This approach also allows the direct effect of low temperature to be studied in isolation from other mortality factors that may be active in the field. Once the survival/mortality ratios are known for various temperatures, biochemical and physiological studies (5) can be included on a logical basis. For instance, HPLC and GLC analyses of antifreeze compounds would not be a priority experiment with insects such as aphids, which die before they freeze, but would be important for species that survived until they froze.

In the conclusion to a review on 'biochemical correlates to cold hardening in insects', Baust (1981) commented that: 'the puzzle [of how an insect survives following a freezing encounter] has many more pieces than we provide by measurements of supercooling points and glycerol content'. Even if we add our assessments of haemolymph freezing points, nucleator content, antifreeze proteins, bound water, DSC and NMR spectra, it is doubtful if the puzzle has become more than half complete. And we know little or nothing about the cause of death in the insects that did not survive!

Few physiologists or biochemists are familiar with the techniques for assessing insect population densities or monitoring microclimate conditions; conversely the expertise of field ecologists will rarely extend to a deep understanding of the metabolic pathways and enzyme kinetics involved in antifreeze production. For these reasons it is almost inevitable that an individual scientist will only be looking at a small part of Baust's large puzzle. We can only marvel at the all-round abilities of the cold-hardy insect, which can locate a favourable overwintering site, reduce its body water content, starve itself and then respond to environmental cues that switch on enzymes to synthesise and regulate cryoprotectants. At times it seems like an unfair challenge.

Note added in proof: The following are reprinted with permission from Pergamon Press Ltd, Headington Hill, Oxford OX3 0BW, UK. *Fig. 4.2: J. Insect Physiol.,* **16**, J. G. Baust & L. K. Miller, Variations in glycerol content and its influence on cold hardiness in the Alaskan carabid beetle *Pterostichus brevicornis,* pp. 979–90, Copyright 1970. *Figs. 4.3 and 4.4: Comp. Biochem. Physiol.,* **73A**, S. Van der Laak, Physiological adaptations to low temperature in freezing tolerant *Phyllodecta laticollis,* pp. 613–20, Copyright 1982. *Fig. 4.5: Comp. Biochem. Physiol.,* **73A**, R. A. Ring, Freezing tolerant insects with low supercooling points, pp. 605–12, Copyright 1982. *Fig. 4.8:* from *Comp. Biochem. Physiol.,* **73A**, K. E. Zachariassen, Nucleating agents in cold-hardy insects, pp. 557–62, Copyright 1982. *Fig. 4.10: Comp. Biochem. Physiol.,* **73A**, J. G. Dunman *et al.,* Antifreeze agents of terrestrial arthropods, pp. 545–55, Copyright 1982. *Figs. 4.13 and 4.14: J. Insect Physiol.,* **32**, M. R. Worland & W. Block, Survival and water loss in some Antarctic arthropods, pp. 579–84, Copyright 1986. *Fig. 4.15: Comp. Biochem. Physiol.,* **73A**, J. G. Baust, Environmental triggers to cold hardening, pp. 563–70, Copyright 1982.

5

Costs and benefits of overwintering

> Winter hibernation in insects seems to be a relatively recent acquisition since the appearance of the glacial climate
> *Levins (1969)*

Introduction

The statement above explains to some extent, the facts that we reveal in this chapter; that overwintering is costly in terms of survival and that the strategies used to overcome these costs by different insect species vary greatly, not only between families but within families and even within species. Moreover, it emphasises the fact that the resources used are very similar but that the way they are used is dissimilar.

When discussing the costs and benefits of the overwintering habit it must be remembered that we are looking at the subject from two slightly different viewpoints. First, we are comparing the overwintering habit, i.e. the ways in which winters are passed, of temperate insects with that of tropical insects, which do not have a dormant phase. Secondly, we compare those insects from temperate climates that do possess a dormant phase with those that do not. Within this category we are also comparing the different overwintering stages of those insects that have opted for a mixed strategy option, i.e. where two or more different overwintering forms are present not only in the same species but in the same location. By adopting these different viewpoints we are able to discuss the advantages and disadvantages of overwintering in the one chapter.

Developmental stage and overwintering success

Insects can pass the winter in a number of different stages – egg, larva, adult and pupa – in fact every stage of the life cycle is used by some species of insect in which to overwinter. Some stages are considered more suitable than others, and not necessarily with respect to cold-hardiness or apparency. For example, it is considered to be an advantage for pest species such as the gypsy moth, *Lymantria dispar* (L.) (Lepidoptera: Lymantriidae), and the moth *Malacosoma disstria* (Hubner) (Lepidoptera: Lasiocampidae) to overwinter as eggs, as this improves synchrony with bud burst of their host plants (Nothnagle and Schultz 1987). However, as well as passing the winter in a number of possible developmental stages, an insect has the option of passing the winter either as an inactive (usually dormant or quiescent) form or as an active one. The adaptations required to pass the winter in such startlingly different manners are, as one would expect, totally different in the approach to winter (see Chapter 3) and the behavioural responses during the winter, but chemically they tend to be very similar (see Chapter 4). A third manner of overwintering is where the insect remains inactive but basically avoids winter, or at least the winter conditions, i.e. cold (winter) avoidance. Each of these three strategies of surviving winter depends very much on the stage of development of the insect at the onset of the usually harsh winter climate.

Winter inactive insects

The developmental stage at which an insect approaches and passes the winter is of great importance to that individual insect and its potential offspring. Depending on the severity of the winter it could literally mean the difference between life and death. At first glance the solution would appear simple – pass the winter safely as a cold-resistant egg hidden in the crevices of tree bark, or avoid the effects of winter cold by pupating deep in the soil. Certainly, these are two tactics that are widely used by insects but even in Finland, where winter temperatures of $-30\,°C$ are common, some members of the Lepidoptera pass the winter as adults, larvae, eggs and pupae (Table 5.1; Seppanen 1969). However, only 19 per cent of them exhibit a mixed overwintering strategy, while 81 per cent overwinter in only one stage of their life cycle. This preponderance of single stage overwintering may be attributable to a generation of insects moving through the year as a single cohort and, unless more than one generation or overlapping generations are present, then a single over-wintering form results. However, this is not always the case. It is obvious that in some species at least, the insect is specially adapted to produce more than one overwintering stage from the same cohort. For example, the speckled wood butterfly, *Pararge aegeria* (L.) (Lepidoptera: Satyridae), had always been considered to produce its mixed overwintering strategy as a result of overlapping broods. However, Lees and Tilley (1980) found that within a single brood the larvae, although hatching at the same

Table 5.1. *Overwintering stages of Finnish macrolepidoptera (data from Seppanen 1969). (Species occur more than once under the mixed strategies column)*

Overwintering stage	No of species using one strategy only	No. of species using mixed strategies
Egg	115	9
Egg–larvae	12	6
Small larvae	92	92
Medium larvae	113	111
Large larvae	48	125
Pupae	277	29
Developed pupae	33	8
Adult	28	3
Total no. of species	718	165

time and kept under identical conditions, all grew at different rates, some going into diapause at the third instar whilst others grew on to pupae and diapaused at that stage. Lees and Tilley (1980) postulated that this was the result of a complex interaction between photoperiod and temperature, dependent on the time of egg-laying, but although this explains why some broods behave differently from others, i.e. producing a single overwintering stage, it does not explain why individuals within the same brood respond differently to identical conditions.

Interestingly, the pupae of the speckled wood butterfly remain in diapause and are cold-tolerant, whereas the late winter larvae become active and feed on grasses during warmer periods of the winter. The loss of some larvae through exposure to cold weather is affordable, because of the high pupal survival, but in mild winters the risk pays off, as the larvae mature early and get an early start to the season. This versatility allows *P. aegeria* to appear as an adult in early spring and also late into the autumn in milder climates (Fig. 5.1). It is perhaps this versatility that allowed *P. aegeria* to extend its range in the British Isles during the middle of this century (Ford 1945) and in north-east Scotland over the last 20 years (Barbour, 1986).

Familial constraints

It would appear that within the Lepidoptera the use of a particular overwintering stage is more of an ancient familial characteristic than a

Table 5.2. *Overwintering stages of some British moths (Lepidoptera)*

Stage	Family: numbers of species				
	Lasiocampidae	Sphingidae	Arctiidae	Notodontidae	Total
Egg	5	0	0	0	5
Small larvae	2	0	9	0	11
Medium larvae	0	0	7	0	7
Late larvae	2	0	9	0	11
Pupae	2	12	8	23	45
Adult	0	2	0	0	2
Total	11	14	33	23	81

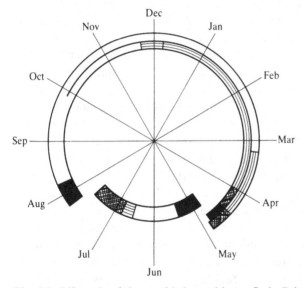

Fig. 5.1. Life cycle of the speckled wood butterfly in Britain showing how mixed strategy overwintering arises (■ egg, □ larvae, ▤ pupae, ▨ adults).

recently evolved response associated with the severity and duration of winter. For example, if four important British moth families are considered together (Table 5.2), each stage of the life cycle is represented, but when looked at separately it is obvious that each family has a stage that predominates, i.e. pupae in the Notodontidae and Sphingidae, larvae in the Arctiidae and eggs in the Lasiocampidae. It is interesting to note that 55 per cent of the overwintering forms are pupae. This familial tendency also appears in butterflies; for example, the Papilionidae and Pieridae mostly overwinter as pupae or adults whilst the Satyridae overwinter as eggs or larvae (Wiklund 1975). Ford (1945) considered 63 different British butterfly species and found that 14 per cent overwintered as eggs, 56 per cent as larvae, 17 per cent as pupae and 11 per cent as adults. The remaining 2 per cent overwintered as larvae or pupae. A similar picture is seen in North American Lepidoptera. This difference between butterflies and moths is interesting, and also indicative of a familial constraint in overwintering stage rather than in a limitation imposed by habitat.

Constraints imposed by life style

Although, as demonstrated above, familial constraints do determine to a certain extent the development stage at which Lepidoptera overwinter, this does not mean that their overwintering strategies are totally indepen- dent of habitat constraints. For example, and once again using the Lepidoptera of Finland but this time confining our attentions to the butterflies (Rhopalocera), we find that although the majority of Finnish butterflies overwinter as cocoons (pupae) there are distinct latitudinal trends in overwintering strategy (Fig. 5.2). Mikkola (1980) analysed these trends in detail and found that overwintering in the adult stage was confined entirely to the most southern part of Finland and that, surpris- ingly, overwintering as a larva was characteristic of those butterflies found in the northern half of Finland. Overwintering as the egg stage was also found to be a southern characteristic, whereas butterflies in central Finland seemed to favour overwintering in the pupal stage.

In high-latitude habitats, development often lasts more than 1 year due to the short growing season, and larval feeding stages persist throughout the winter. A particular insect may often pass more than one winter as a larva, e.g. Tipulidae (Diptera) (MacLean 1973). Many beetles living in wood, e.g. Cerambycidae, have life cycles of more than 3 years and also overwinter as larvae through more than one winter. In the severe arctic conditions existing in Ellesmere Island, 92 per cent of the 90 species of insect in which the overwintering stage has been discovered overwinter as larvae (Danks and Byers 1972).

The preponderance of a particular developmental stage as the over- wintering stage, in the case of the macro-Lepidoptera the pupa or cocoon, is often related to the life style of the insect. Most aquatic insects overwinter as larvae (Danks 1978), whereas terrestrial insects exhibit a variety of overwintering stages, but clearly linked to their overall familial

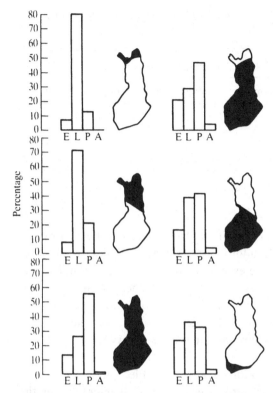

Fig. 5.2. Proportion of butterflies overwintering in different stages throughout Finland. E, eggs; L, larvae; P, pupae/cocoons; A, adults (redrawn after Mikkola 1980).

strategy. For example, there are at least 214 species of European ichneumon-flies (Hymenoptera: Ichneumonidae) that overwinter in the adult stage (Rasnitsyn 1966), whereas in North America out of a total of 13 genera from three subfamilies only 39 species of ichneumonid overwinter as adults, and these are mainly female (Dasch 1971). Generally, the stage is more or less the same within a genus but seldom within a family, although as noted above one stage often predominates, e.g. most Miridae overwinter as eggs. Familial characteristics are strikingly revealed in a study by Hodson (1937) on the overwintering strategies of important pest species of North America (Table 5.3). Homoptera, for example, overwinter mainly in the egg stage, whereas Heteroptera overwinter mainly as adults.

Aquatic insects generally overwinter as larval stages; their pupae are often susceptible to changes in the environment and the aerial adult forms

Table 5.3. *Overwintering stages of important North American insect pests* (*after Hodson 1937*)

Stage	Order: percentage of species*				
	Lepidoptera	Coleoptera	Homoptera	Hemiptera	Diptera
Egg	11.7	2.5	75.0	25.0	50.0
Larvae	45.1	47.5	16.6	0	0
Pupae	43.1	0	–	–	37.5
Adult	7.8	65.0	25.0	75.0	12.4

* Percentages do not add up to 100 as some species exhibit mixed overwintering strategies.

would be exposed to harsher conditions than those experienced by the aquatic larvae or nymphs. In addition, the larvae generally live close to, or in a sheltered substrate. The exceptions to these are those individuals that conceal overwintering eggs in, or firmly attached to, the substrate. Larvae, being mobile, can actively seek out overwintering sites by burrowing, whereas eggs remain in the sites where they were laid. Danks (1978) cites the following groups of aquatic insects with overwintering larvae: Chironomidae, many Simuliidae, most Trichoptera, Megaloptera, some Ephemeroptera, most Odonata and most Plecoptera. Exceptions are found where the adult is also aquatic, e.g. Coleoptera, or in the Culicidae, where the larvae are floaters and cannot burrow for protection.

Non-feeding and gut evacuation

Harsh conditions may require a specialist stage for overwintering which is either non-feeding in nature (egg and pupa) or evacuates the gut in preparation for winter (larvae and adults). Food particles in the digestive system can act as sites for ice nucleation and reduce the inherent ability of the insect to supercool (Salt 1953). For instance, supercooling in the beech weevil, *Rhynchaenus fagi* (L.) (Coleoptera: Curculionidae), is markedly reduced when feeding resumes in spring (Bale 1980). There is a slight loss in cold-hardiness (supercooling) in first instar aphids after the initial imbibition of plant sap, but conversely the extensive super-cooling observed in aphids that overwinter anholocyclically may be an inadvertent but largely irrelevant byproduct of antifreeze protection gained from the high solute concentration of the phloem, as aphids die before the freeze (O'Doherty and Bale 1985; Knight and Bale 1986).

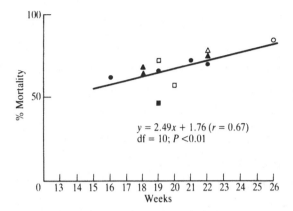

Fig. 5.3. Relationship between length of winter weeks (i.e. time for egg deposition to first egg hatch) and egg mortality in aphids. ● *Rhopalosiphum padi* (Scotland), ■ *R. padi* (England), ○ *R. padi* (Finland), □ *Drepanosiphum plantanoidis* (Scotland), ▲ *Euceraphis punctipennis* (Scotland), △ *Tuberolachnus annulatus* (Scotland).

Overwintering mortality and population regulation

In many insects, overwintering mortality is of great importance in controlling population levels, e.g. the pine beauty moth, *Panolis flammea* (D & S) (Lepidoptera: Noctuidae), can suffer up to 40 per cent pupal mortality during the winter (Leather 1984). This source of mortality is regarded as one of the more important factors in keeping *P. flammea* populations below outbreak levels in certain forest sites (Walsh 1990). A number of aphid species have been shown to suffer overwintering egg mortality in excess of 50 per cent (Leather 1983; Kidd and Tozer 1985), with more than 90 per cent killed in the active stages (adults and nymphs) (Harrington and Cheng 1984; Hand and Hand 1986; Knight and Bale 1986). In fact, even in the supposedly resistant egg stage, overwintering mortalities in excess of 70 per cent are the norm rather than the exception (Leather 1990). Although in some cases overwintering egg mortality of aphids cannot be strictly regarded as a regulatory factor, the length of the winter can markedly influence the number of eggs surviving (Fig. 5.3) and thus the initial success of the aphid population in the spring. This is certainly the case in the bird cherry-oat aphid, *Rhopalosiphum padi*, and the black bean aphid, *Aphis fabae* (Homoptera: Aphididae), where the initial hatching population is closely linked with the subsequent success of the population (Way *et al.* 1981; Leather 1986).

In some insects, overwintering mortality has been shown to be not only an important factor but, as in the case of the adelgid *Gillettellia cooleyi*

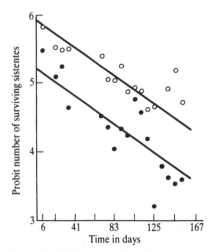

Fig. 5.4. The proportion of sistentes remaining alive in (○) young colonies and (●) old colonies of *Gillettellia cooleyi* sistentes on Douglas fir needles (after Parry 1980).

(= *Adelges cooleyi*) (Gillette) (Homoptera: Adelgidae), the key factor controlling population levels. This insect lives on Douglas fir, *Pseudotsuga menziesii*, in Scotland and can enter the winter as either an old colony, made up of adults and late instar nymphs, or as a young colony composed of newly born and early instar nymphs (Parry 1978). Parry (1980) collected sample twigs from Douglas fir during two winters and found that the number of *G. cooleyi* individuals declined linearly over the winter period (Fig. 5.4). There were significant differences between the two differently aged colonies, mortality being greater in the old colony than in the young colony, although this difference was attributed to the differences in early winter (prediapause) mortality, as the slopes for post-diapause overwintering were identical. It has been noticed that feeding can make insects more vulnerable to freezing by reducing their ability to supercool (Salt 1953) and this is seen clearly in Parry's study, where mortality from low temperatures increased dramatically once feeding commenced (Fig. 5.5). It is thus very important for *G. cooleyi* to enter the winter as a non-feeding stage.

Winter active insects

In some habitats, such as timber or streams, growth continues during the winter, e.g. blackfly larvae (Diptera: Simuliidae). In the Arctic carabid,

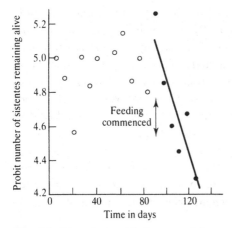

Fig. 5.5. The relationship between time, commencement of feeding and proportion of sistentes remaining alive on Douglas fir needles (after Parry 1980).

P. brevicornis, feeding begins after January and egg development commences in mid-February, although the thaw does not occur until May (Kaufmann 1971).

Other insects are not only capable of feeding but are able to reproduce at extremely low temperatures; these are the real winter active insects. These insects balance on the edge of their survival limits. For example, the snow fleas, *Boreus hyemalis* (L.) and *B. westwoodi* (L.) (Mecoptera: Boreidae), and the crane fly, *Chionea araneoides* Dalm. (Diptera: Tipulidae), have supercooling points of −5.7, −5.6 and −7.5 °C, respectively (Sømme and Ostbye 1969). Although air temperatures in Norway frequently drop below these lethal temperatures, the insects can remain active in a supercooled state, and by keeping to the subnivean layers whilst temperatures are below their supercooling points, they are able to survive.

The southern corn rootworm, *Diabrotica unidecimpunctata howardi* Barber (Coleoptera: Chrysomelidae), mates at the beginning of winter. As males rarely survive winter, it is important that mating is accomplished as quickly as possible. Both males and females elevate their body temperatures above ambient air temperature by basking in direct sunlight. This habit provides males with extended daily time periods to locate and mate with females and later in the winter enables postreproductive–diapause females to feed and to mature eggs, ready to exploit fresh foliage in the spring (Meinke and Gould 1987).

Courtin *et al.* (1984) were intrigued by the ability of the snow scorpion fly, *B. brumalis*, which is considered to be the most abundant and widespread species of snow scorpion fly in North America, to remain active throughout winter. This is a small, black, flightless insect with long black legs and antennae. It remains active on the surface of the snow during the day, gaining enough energy from shortwave radiation from the sun both from above and, by virtue of its long legs, from reflection from the surface of the snow. Also as a result of its long legs, heat loss by conductance to the snow is minimised while at the same time its body is still within the warmer boundary layer. On less favourable days the snow scorpion fly remains beneath the snow, where conditions are less harsh. Shorthouse *et al.* (1980) found that at depths of 1 m the temperature was never lower than $-3\,°C$, even when surface temperatures of $-25\,°C$ were experienced.

Other insects that would normally not be considered winter active, e.g. the aphid *Brachycaudus helichrysi* (Mordv.) (Homoptera: Aphididae), also show similar adaptations to survive the winter (Bell 1983). The nymphs are able to survive temperatures of $-15\,°C$ and, although eggs are laid at the beginning of winter, they usually hatch in mid-February. This is a direct result of their positioning; they are laid at the top of the bush and in warm conditions they have been known to hatch as early as November. On hatching, the aphids feed on the side of the bud nearest to the stem but oriented towards the sun; presumably this location and positioning provides a local microclimate with a higher temperature than that of the surrounding air.

Winter avoidance

As well as having stages of the life cycle that pass the winter as dormant or quiescent forms, a number of insect species can also pass the winter successfully in what would normally be considered their active summer stage. These winter active species are characterised by the ability to seek out less harsh environments and are thus able to exploit a different number of stages and increase their growth and developmental rates in comparison with those species that halt development during winter.

The collembolans *Isotoma hiemalis* (Schott) (Isotomidae), *Entomobrya multifasciata* (= *nivalis*) (Tullberg) (Entomobryidae) and *Hypogastrura socialis* (Uzel) (Hypogastruridae) are not freezing-tolerant and can be killed at temperatures as high as $-10\,°C$, yet by migrating to the subnivean layers are able to survive by avoiding temperatures as low as $-25\,°C$ (Sømme 1976). The large elm beetle, *Scolytus scolytus* (F)

(Coleoptera: Scolytidae), overwinters as a larva and feeds throughout the winter, although as it lives within the tree, i.e. a highly protected environment, it is debatable whether this is a real winter active species or not (Barson 1974). This is probably true of most sub-bark-feeding beetles, even though many species, e.g. *Ips acuminatus*, Gyll (Coleoptera: Scolytidae) possess all the cold-hardiness attributes of animals living in more exposed situations. In the case of *I. acuminatus*, however, as it hibernates under the thin bark of Scots pine, *Pinus sylvestris*, (Gehrken 1984) it is probably more an overwinterer than a winter avoider.

In extreme climates, such as Scandinavia and Iceland, some interesting adaptations for overwintering success have arisen. For example, house and stable flies, *Musca domestica* L. (Diptera: Muscidae) and *Stomoxys calcitrans* (L.) (Diptera: Muscidae), survive the extremely cold winters of Norway as fully active breeding adults and larvae (Sømme 1961). This is achieved by the simple expedient of passing the winter in animal houses, which, due to their solid construction and large amounts of animal droppings, quite commonly maintain temperatures at 15–18 °C.

The bird cherry-oat aphid, *Rhopalosiphum padi*, which possesses a number of ways of overwintering – cold-hardy inactive egg stages, cold-hardy active stages – also possesses the avoidance option, albeit somewhat fortuitously. In Canada, nymphs and adults of *R. padi* have been found feeding on germinating wheat seeds in grain stores (Harper and Blakeley 1968) and in Iceland, where the usual overwintering host, *Prunus padus*, is absent, they have been found feeding on grasses by the side of hot water springs (Hille Ris Lambers 1955).

The weevil *Strophosomus melanogrammus* (Forster) (Coleoptera: Curculionidae), a pest of Sitka spruce, *Picea sitchensis*, in forest nurseries in Scotland (Parry 1981) overwinters as an adult. It remains active and feeds throughout the year. However, the supercooling points of this insect are only in the range of -3.7 to -4.5 °C and as air temperatures in Scotland frequently fall below this range, avoiding action must be taken. On frosty days, the beetles take advantage of the vertical stratification of temperature in the soil (see Chapter 2). Adult *S. melanogrammus* drop from the foliage and then, once on the soil surface, where temperatures are even lower, burrow into the soil and escape the rigours of above-ground conditions (Fig. 5.6). Their respiration rates also fall during the winter, especially when they are underground and have no available food (Parry 1983). This insect is a true winter avoider in that it is opportunistic and not committed to any one overwintering form or site.

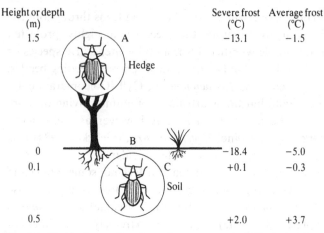

Height or depth (m)		Severe frost (°C)	Average frost (°C)
1.5	A, Hedge	−13.1	−1.5
0	B	−18.4	−5.0
0.1	C, Soil	+0.1	−0.3
0.5		+2.0	+3.7

Fig. 5.6. Vertical stratification of temperature (°C) on nights with severe and average frosts, showing how the weevil *Strophosomus melanogrammus*, avoids lethal temperatures by changing its overwintering site. A, feeding site on foliage during warmer winter weather; B, vegetation layer at soil surface where unable to survive; and C, resting site during frost periods.

Some insects, although seemingly extremely well protected to survive severe overwintering conditions, rarely seem to experience them. For example, the cabbage root fly, *Delia radicum* (L) (Diptera: Anthomyiidae), overwinters as a pupa within a puparium, in the soil around the roots of brassica crops. In southern England, they habitually overwinter at depths of 2.5–7.5 cm below the soil surface. During the winter of 1983–4 a population was studied (Block *et al.* 1987) near Ascot. At no time in winter was the soil temperature below 0 °C, with a mean of 3.1 °C. Yet overwintering pupae of *D. radicum* have a mean supercooling point of −20 °C. This seems a clear case of an insect with a perfectly adequate overwintering system avoiding the issue by spatial distribution in winter. It will be interesting to see if populations of *D. radicum* in these and similar climates lose their supercooling powers. However, as many tropical insects still retain supercooling abilities (Cloudsley-Thompson 1973) this may never occur.

In summary, it appears that the stage used for overwintering is correlated with the taxon of the insect and is influenced by general life cycle features and the severity of the conditions experienced or likely to be experienced. This in turn depends on latitude and habitat, which are discussed in Chapter 2.

The costs of overwintering

The rather broad questions addressed in this section are what are the costs of overwintering? and are they any more 'expensive' than the costs incurred during the rest of the year? The first part of this question will be answered in the following pages of this section, but the rather more speculative analysis of the cost:benefit ratios between winter and the rest of the year will be dealt with at the end of the chapter.

That there are costs to overwintering is axiomatic, how the costs compare between strategies is the debatable point. Mansingh (1971), when defining hibernation as: 'a physiological condition of growth retardation or arrest, primarily designed to overcome lower than optimum temperatures during winter . . .' added that: 'almost always the overwintering insects have also to face several other adversities associated with winter conditions.'

The costs of overwintering can be, for the sake of simplicity, ascribed to two causes – the costs incurred by virtue of being non-mobile, i.e. predation and upredictable events, and the costs incurred by other factors, e.g. desiccation, cold injury, etc. The fact that these two costs exist and act together was demonstrated by Leather (1981), in a study of overwintering eggs of the bird cherry-oat aphid, *Rhopalosiphum padi*. The eggs, which were laid in the axils of buds on young twigs of the bird cherry, *Prunus padus*, were protected from (i) arthropod predators; (ii) all predators; and (iii) exposed to all predators and the weather. The differences in mortality were very marked; only 35 per cent mortality occurred in the total predator exclusion treatment, 66 per cent mortality in the arthropod predator exclusion treatment with 81 per cent mortality was seen in the control treatment (Fig. 5.7). There were thus demonstrated to be two major causes of mortality – predation and natural causes/weather. This joint cost to the overwintering stages of aphids is seen in many other species, e.g. *Aphis fabae* Scop., *Acyrthosiphon pisum* (Harris), *Aphis pomi* De Geer and *Schizolachnus piniradiatae* (Davidson), which show overwintering mortality rates of 40, 83, 75 and 70–90 per cent, respectively (Peterson 1917; Dunn and Wright 1955; Way and Banks 1964; Kulman 1967) (Table 5.4). This mortality in *R. padi* is a function of the time spent in the overwintering stage, i.e. the longer the winter lasts the greater the mortality, which occurs in a log-linear fashion (Leather 1980a, 1981; Leather and Lehti 1981). This relationship not only seems to apply to other species of aphids, e.g. *Drepanosiphum platanoidis* Schrank, *Euceraphis punctipennis* Zett. and *Tuberculoides annulatus* (Hartig), but

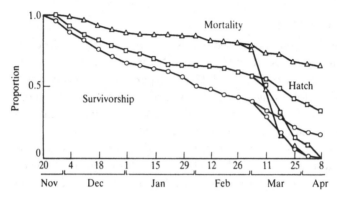

Fig. 5.7. Egg survival of *Rhopalosiphum padi* on *Prunus padus*. ○—○ control, □—□ partial predator exclusion, △—△ total predator exclusion (after Leather 1981).

is also seen from country to country (see Fig. 5.3). High overwintering egg mortality is not confined to aphids; mortality of gypsy moth, *Lymantria dispar* (L.) (Lepidoptera: Lymantriidae), eggs can be as high as 85 per cent (Leonard 1972), mortality of eggs of the praying mantis, *Mantis religiosa* L. (Orthoptera: Mantidae), ranges from 15–86 per cent, depending on weather (Salt and James 1947) and eggs of the pine sawfly, *Neodiprion sertifer* (Geoff.) (Hymenoptera: Diprionidae), suffer 52 per cent overwintering mortality in Norway (Austarä 1971) and mortalities ranging from 7–96 per cent have been reported in Canada for the same

Table 5.4. *Total mortalities suffered by the overwintering eggs of different species of aphid*

Species	% mortality	Authority
Acyrthosiphon pisum	50–67	Bronson (1935)
Acyrthosiphon pisum	83	Dunn and Wright (1955)
Acyrthosiphon pisum	72	Bournoville (1973)
Aphis avenae	40–70	Peterson (1920)
Aphis fabae	40	Way and Banks (1964)
Aphis pomi	75	Peterson (1917)
Cinara pinea	50	Kidd and Tozer (1985)
Pterocallis alni	65	Gange and Llewellyn (1988)
Rhopalosiphum padi	70	Leather (1980a)
Rhopalosiphum padi	80	Leather (1981)
Schizolachnus piniradiate	78	Grobler (1962)
Schizolachnus piniradiate	70–90	Kulman (1967)

species (Sullivan 1965). Thus, there would appear to be a specific cost attached to the duration of the overwintering stage in aphids, those in more arctic conditions being at a disadvantage compared to the temperate regions. The duration of the overwintering stage also affects the survival of the adult boll weevil, *Anthonomus grandis* Boheman (Coleoptera: Curculionidae), in Texas, where less than 10 per cent habitually survive the winter months (Fuchs and England 1989).

Spreading the risk

In endopterygote insects, larvae and pupae may be favoured as over-wintering stages because of their greater haemolymph volume which increases freezing resistance (Asahina 1966, 1969). Where insects are permanent inhabitants of stable environments such as the soil, which provides a physical barrier to cold, there is no selective advantage to overwinter in a particular stage, although for other reasons (e.g. food availability) a single morph may predominate (see p. 152).

Nevertheless, some of the most resistant species overwinter in several stages, e.g. the Antarctic springtail, *Isotoma klovstadi* (Carpenter) (Collembola: Isotomidae), although this does usually overwinter as an egg (Pryor 1962). The Arctic carabid, *Pterostichus brevicornis* (Kirby) (Coleoptera: Carabidae), on the other hand, overwinters in all stages except the egg (Kaufmann 1971). In some species of insect the less hardy stages are able to overwinter in milder than normal winters or in regions with a more favourable climate, e.g. the overwintering of various morphs of aphid species (Woodford and Lerman 1974; Williams 1980). In New Zealand the introduced German wasp, *Vespula germanica* (Fabricus) (Hymenoptera: Vespidae), is able to overwinter in most years as whole colonies rather than as individual queens and extremely large colonies result (Thomas 1960), whereas in Britain only individual queens survive the winter and large colonies are the exception rather than the rule.

Mixed strategies and variable costs

Aphids also provide us with a good example of a mixed overwintering strategy. The bird cherry-oat aphid, *Rhopalosiphum padi*, can overwinter both as the egg form and as a viviparous adult or nymph. During the winters of 1977–8 and 1978–9, populations of the former were followed on bird cherry trees and of the latter on various grasses in East Anglia. In both years the mortality of those aphids overwintering as active forms was 100 per cent (Leather 1980a). There is no doubt as to the disadvantage of overwintering as an active stage in this case. However, in other parts of England, *R. padi* can and does overwinter successfully as the active stage (Williams 1980).

A mixed strategy of overwintering is clearly advantageous for aphids. There is an inescapable mortality factor of up to 80 per cent in the egg stage but a guarantee of survival of some eggs, whereas adults or nymphs, although less cold-tolerant, may survive and continue to reproduce in midwinter (Leather 1980a). Thus, to overwinter wholly as an egg sacrifices a possible population increase in favour of a definite decrease in population size. In addition, if overwintering has occurred as an egg, the population cannot begin to develop until the egg has hatched and the first aphids have reached adulthood; those aphids overwintering as adults can begin to initiate a new population much earlier. In this case, climate is the major factor determining the choice of overwintering strategy. A similar example of a mixed overwintering strategy is seen in water striders (Heteroptera: Gerridae), which exist in two morphs – long wing and short wing – the latter being more cold-susceptible than the former. The frequency of the two forms varies across Europe, only long winged forms being found in Finland, where winters are extreme, whereas mixed populations are found further south in Europe (Vepsäläinen 1974).

Some insects overwinter more than once in their life cycle. Antarctic collembola for example can take from 2 to 3 years to reach maturity and the moth, *Gynaephora groenlandica* (Lepidoptera: Lymantriidae), takes 10–14 years (Kukal and Kevan 1987). These insects are thus forced to adopt a mixed strategy. The collembolan, *Parisotoma octooculata* (Collembola; Isotomidae), overwinters as either the egg or late larval stages. Another Antarctic collembolan, *Cryptopygus antarcticus* (Collembola: Isotomidae), has many overlapping generations and overwinters at each of these stages (Burn 1984). Their ability to withstand winter temperatures varies, *C. antarcticus* is better able to withstand the winter and has adopted a long life cycle with continuous reproduction throughout the summer. In contrast, *P. octooculata* is less able to survive in winter but avoids repeated overwintering by maturing faster. Thus, most over-wintering is done by the most cold-hardy stage – the egg. This strategy requires synchronous reproduction and the population would be at risk if a catastrophe occurred in the spring. So, this latter strategy may have more costs associated with it, in the long run, than the apparently high cost strategy adopted by *C. antarcticus*.

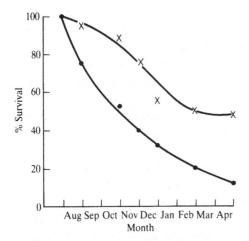

Fig. 5.8. Survival of the pine shoot moth, *Rhyacionia buoliana*, in Connecticut, USA. ●—● during the severe winter of 1931–2, ×—× during the mild winter of 1932–3 (after West 1936).

High mortality rates seem to be a characteristic of overwintering insects irrespective of stage, e.g. the codling moth, *Cydia pomonella* (L.) (Lepidoptera: Tortricidae), which overwinters as a larva, suffers over 80 per cent mortality during a typical winter (Causse 1976). Another tortricid, the pine shoot moth, *Rhyacionia buoliana* (D & S), also shows mortality rates of 63–92 per cent, depending on the severity of winter experienced (West 1936). In a mild winter the pattern of mortality is completely different from that seen in the severe winter (Fig. 5.8). Pupal mortality, on the other hand, can be fairly low in some Lepidoptera, e.g. 27–40 per cent in *Stilbosis quadricustatella* (Cham.) (Lepidoptera: Cosmopterigidae) (Mopper *et al.* 1985) and 35–45 per cent in the pine beauty moth, *Panolis flammea* (D & S) (Lepidoptera: Noctuidae) (Leather 1984).

Physical costs

The costs of overwintering are not solely attributable to the severity of the winter, for example, many insects, e.g. the pine beauty moth, *Panolis flammea*, overwinter as pupae in the soil. Temperatures are less severe further down in the soil than nearer the surface (see Chapter 2) and there is thus an advantage to be gained by burrowing. However, deeper burrowing does have some disadvantages. There are at least three costs associated with burrowing: (i) the more effort spent on burrowing, the less reserves remaining with which to overwinter; (ii) spring comes later to deep layers in the same way that deeper layers remain warmer, so that a cost is incurred by emerging later and perhaps failing to mate; and (iii) spring emergence is usually heralded by a change in form, from a hard pupae to a soft-bodied pharate adult. If the insect has burrowed too deeply, then by the time the adult reaches the surface, if it is able to do so, its wings may be torn, so that flight, and thus a successful mating, is impossible.

Perhaps an even more important factor is the relative saving to be made in terms of energy consumption or resource allocation to a particular mode of overwintering or overwintering site. One of the most elegant pieces of work to address this particular aspect of overwintering is that of Hoshikawa *et al.* (1988), who demonstrated how the overwintering strategies of eleven species of insect from a number of different families approached the problem common to them all – how to survive winter. In essence, those insects such as the beetle, *Popillia japonica* (Coleoptera: Scarabaeidae) that overwinter at depths of more than 10 cm below the surface have higher supercooling points and different overwintering strategies from insects such as the spotted cutworm, *Xestia c-nigrum* (Lepidoptera: Noctuidae), which overwinters at the soil surface and has a lower lethal temperature of $-20\,°C$ compared to that of *P. japonica*, which is only $-6\,°C$. The beetle *Gastrophysa atrocyanea* (Coleoptera: Chrysomelidae), which overwinters just below the soil surface has, as expected, an intermediate lower lethal limit of $-10\,°C$ (Fig. 5.9). Thus, those species overwintering at greater depths have less need to use valuable reserves in the expensive production of cryoprotectants. In addition, insects overwintering at these greater depths are able to remain active at quite low temperatures due to their energy-saving strategy; larvae of the march fly, *Bibio rufiventris* (Diptera: Bibionidae), can escape into still deeper layers should lower winter temperatures occur. Three strategies of overwintering were pinpointed: (i) the supercoolers; (ii) the frost resistors; and (iii) the cold avoiders. The frost resistors as would be expected overwinter closer to the soil surface.

Metabolic costs

Before successful overwintering can occur, insects must prepare for the potentially dangerous conditions that lie ahead; these biochemical changes

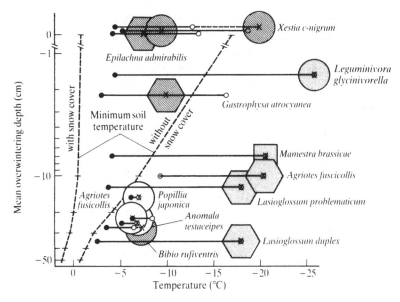

Fig. 5.9. Overwintering depths in soil and overwintering strategies for 11 insect species in Japan. Supercooling points (○), inoculative freezing points (●) and lower lethal temperature, (×) are given for each species. Three strategy types, supercoolers (▦), frost resistors (▨) and cold avoiders (□) are shown at the lethal temperatures for larvae (large circles), pupae (squares) and adults (hexagons). The annual minimum temperatures in snowy and snowless conditions are shown by two dotted lines (after Hoshikawa *et al.* 1988).

associated with cold-hardiness are often substantial, e.g. 25 per cent of the fresh weight of overwintering *Bracon cephi* (Hymenoptera: Braconidae) is glycerol (Salt 1959). We would therefore expect that the maintenance of cold-hardiness would be energy-dependent and depleting.

As Danks (1978) points out, the levels of cold-hardiness seen in different insects are correlated to the severity of the climate in which they live. There are many examples from different insect orders to illustrate this point (see Chapter 4) and this would imply that the unnecessary maintenance of high levels of cold-hardiness is disadvantageous, presumably due to the metabolic costs. For instance, cryoprotectants are usually manufactured only as winter approaches (Sømme 1965b; Miller 1969) and the level of these cryoprotectants can be adjusted throughout the winter depending on the conditions experienced (Table 5.5; Hansen 1973; Eguagie 1974).

Table 5.5. *Changes in the body composition of hibernating females of* Tingis ampliata *during the winter of 1965–6 (after Eguagie 1972)*

Date	Stage in life cycle	Live weight (mg)	Dry weight (mg)	% water	% fat	% non-fat solids
2.10.65	Just before diapause	38.12	10.63	71.8	17.3	11.0
2.11.65	Hibernation and reproductive diapause	31.90	10.14	68.2	16.5	15.3
2.1.66	Hibernation and reproductive diapause	31.60	9.00	70.8	12.7	16.6
2.3.66	Hibernation and reproductive diapause	30.99	13.00	58.1	6.5	35.5

Glycogen

Few studies have attempted to calculate the actual amount of 'fuel' utilised by overwintering insects. However, Rickards *et al.* (1987) working with larvae of the goldenrod gall moth, *Epiblema scudderiana* (Clemens) (Lepidoptera: Olethreutidae), calculated that over the period from mid-October to 4 January, larvae used glycogen at a steady rate of 16.03 µmol/g dry weight (expressed in glucose units), by which date their glycogen reserves were almost exhausted (Fig. 5.10). This utilisation of winter 'fuel' was initiated by the first experience of a subzero temperature and replenishment of the fuel supply did not occur until spring, when temperatures were regularly above the freezing point of water.

These types of indirect evidence support the hypothesis that cold-hardiness is metabolically costly. In fact this energetic cost might limit the range of some organisms physiologically capable of developing cold-hardiness. In addition to the maintenance costs described for the upkeep of cold-hardiness, other metabolic costs are also incurred, for instance, the costs associated with respiration, which are usually measured by calculation of depletion rates of fat reserves within the insect. For example, the monarch butterfly, *Danaus plexippus* L. (Lepidoptera: Danaidae), spends the winter in aggregations of from 35 000–100 000 butterflies in a quiescent state. Even so, the fat content of these butterflies declines from 71 to 36 per cent and indicates a cost of 26.05 joules per day (Chaplin and Wells 1982) (Fig. 5.11).

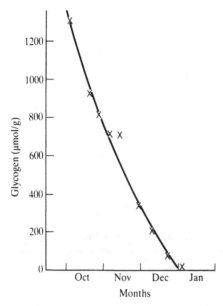

Fig. 5.10. Profiles of glycogen (µmol/g as glucose equivalent) per gram dry weight contents in *Epiblema scudderiana* larvae over the 1984–5 winter (redrawn after Rickards *et al.* 1987).

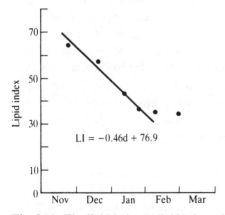

Fig. 5.11. The lipid index (g lipid/g lean dry weight) × 100, of *Danaus plexipppus* butterflies at various times during the winter. The correlation coefficient of the regression of lipid index (LI) and time (days) over the 61-day period from late November to late January is −0.99 (after Chaplin and Wells 1982).

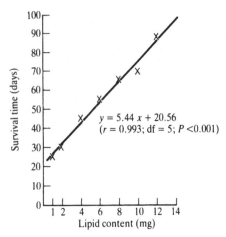

Fig. 5.12. Relationship between survival time and abdominal lipid content in *Inachis io* after 1 day of adult feeding prior to diapause (drawn from data in Pullin 1987).

Lipids

Abdominal lipid contents are vital to the overwintering survival of adult Lepidoptera. This is mediated by the amount of feeding that adults can do before the onset of diapause. Pullin (1987) working on two butterfly species, *Aglais urticae* (L.) and *Inachis io* (L.) (Lepidoptera: Nymphalidae) was able to show significant correlations between lipid content and adult weight. Furthermore, he was able to show that the feeding time allowed to the adults prior to diapause was significantly correlated with the weight of lipids present in the abdomen. Fresh weight declined linearly in both species according to the time spent overwintering. More importantly, there was a significant linear correlation between survival time and initial fresh weight. Therefore the longer adult butterflies fed before diapause the longer they were able to survive over the winter. Using these relationships it is possible to calculate the costs, in terms of lipid, of overwintering in these two species (Fig. 5.12). In the case illustrated it costs *I. io* 0.15 mg of lipid/day whilst overwintering, presumably purely in metabolic costs.

Desiccation

Overwintering insects incur overwintering costs by desiccation as well as by metabolism. For example the predatory anthocorid, *Anthocoris nemorum* (L.) (Hemiptera: Anthocoridae), loses 38 per cent of its fat body and 17.8 per cent of its body weight during the winter (Anderson 1962). The

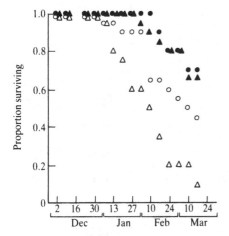

Fig. 5.13. Pupal mortality in male (●, ○) and female (▲, △) *Panolis flammea* in high (●, ▲) and low (○, △) humidity regimes (after Leather 1984).

monarch butterfly, *Danaus plexippus*, in Australia loses 25 per cent of its water content during overwintering as well as 51 per cent of its fat reserves (James 1984). Weight loss can be a critical factor in survival. For example, overwintering pupae of the pine beauty moth, *Panolis flammea*, lose approximately 50 per cent of their wet weight during the winter, and die if their weight falls below 115 mg. Humidity conditions are thus of great importance and under controlled laboratory conditions a three-fold difference in survival rates is seen in pupae kept under moist conditions compared to those kept under dry conditions (Leather 1984) (Fig. 5.13).

Selection pressure

The cold-hardiness of most species does not develop much above the level required for survival, and the cold-hardiness of different geographical races of one species, even under the same conditions, normally differs. Sullivan (1965), working with eggs of the pine sawfly, *Neodiprion sertifer*, laid under identical conditions by adults originating from different parts of the world, was able to show that the cold-hardiness of those eggs arising from populations nearer the poles was greater than those from populations nearer the equator. This would suggest that selection operates against cold-hardiness in areas where such physiological adjustments are not required for survival under local conditions. Sullivan (1965) further postulated that the range of *N. sertifer* would gradually extend through

Table 5.6. *Species and clones of aphids in relation to lower thermal* (LT_{50}) *limits (after Griffiths and Wratten 1979)*

Species	Clone	Winter temperature (°C)	On plant	Off plant (dry)	Off plant (wet)
				LT_{50}	
Sitobion avenae	Spain	11.00	−6.8	−5.00	–
Sitobion avenae	Switzerland	−5.00	–	−10.68	−9.35
Sitobion avenae	Norway	−3.75	−8.8	−8.37	–
Metopolophium dirhodum	Switzerland	−5.00	−8.5	−11.67	−7.55
Rhopalosiphum padi	Norway	−3.75	−8.3	−4.8	–
Rhopalosiphum padi	Finland	−12.00	–	−12.29	−11.96

Canada as selection continued to operate on the introduced population. This phenomenon is also seen in cereal aphids in Europe (Griffiths and Wratten 1979), those aphids from southern Europe having very different lower lethal limits from those from colder regions of Europe. Aphids from Norway were able to survive lower temperatures than aphids of the same species but coming from Spain (Table 5.6).

Reproductive costs

The channelling of resources into preparation for winter can reduce fecundity in diapausing insects. For example, if the codling moth, *Cydia pomonella*, is exposed to short photoperiods during larval development, a shortening daylength being indicative of approaching winter, the adults that emerge from the pupae lay only 50 per cent of the eggs laid by adults that emerge from larvae subjected to long day-lengths (Deseö and Saringer 1975). Furthermore, those individuals that had overwintered had a mean fecundity of about 42 eggs/female, compared to 96 eggs/female shown by the summer generations (Deseö 1973). This is also seen in *Plutella xylostella* L. (Lepidoptera: Yponomeutidae) (Harcourt and Cass 1966) and other Lepidoptera with more than one generation per year (Table 5.7). This trade-off between reproduction and the preparation for winter is particularly exemplified by the bird cherry-oat aphid, *Rhopalosiphum padi* (Ward *et al.* 1984). Aphids depend on asexual parthenogenetic

Table 5.7. *Fecundity of overwintered and summer generations of some Lepidopteran species (after Deseö 1973)*

Species	Mean fecundity		Region
	Overwintered	Summer	
Cydia pomonella (L.)	75	100	CIS
	94	173	USA
	64	83	Canada
	17 (max 79)	22 (max 106)	USA
	21 (max 99)	25 (max 209)	USA
	61	103	USA
Hyphantria cunea Drury	566	799	Hungary
Cydia funebrana (Treit.)	20–85	100–200	Romania
	45	80	CIS
	8	18–59	Hungary
Lobesia botrana (D & S)	77	140	Hungary
Cochylis ambiguella Hubn.	50	73	Hungary

reproduction for the rapid growth of a clone. The production of a resistant overwintering egg requires the production of a sexual generation, which, as it effectively ends the period of rapid growth, should be postponed as late as is compatible with successful egg laying. The time of leaf fall determines the latest possible time of oviposition, and is itself mainly determined by day-length. The time required for the development of the final generations of the aphid depends on temperature. The termination of parthenogenetic generations depends on temperature and photoperiod (Dixon and Glen 1971) and in the field the time of production of male *R. padi* varies adaptively with July temperature (Ward *et al.* 1984) (Fig. 5.14).

Insect colour and overwintering costs

Intraspecific variation in overwintering costs can be seen in many species of insects and some of these have been described earlier in this section, e.g. egg versus active stage in *R. padi* and short wing versus long wing in *Gerris* spp. However, one of the more interesting aspects of this is the different colour morphs of individual insects. In many insects there is a wide variability in the colour forms present within the population, the most well known examples being the two-spot ladybird, *Adalia bipunctata* (L.) (Coleoptora: Coccinellidae), the ten-spot ladybird, *A. decempunctata* (L.) (Coleoptera: Coccinellidae), the English grain aphid, *Sitobion avenae*

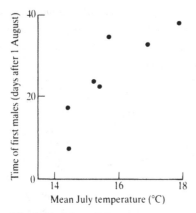

Fig. 5.14. Mean July temperature and the time of first capture of male *Rhopalosiphum padi* in the Brooms Barn suction trap (after Ward *et al.* 1984).

(FO) and the rose aphid, *Macrosiphum rosae* (L) (Homoptera: Aphididae). The two ladybird species have forms ranging from completely black to completely red and the two aphid species range in colour from pale green to dark brown. Colour is an important factor in thermoregulation – light colours acting as heat reflectants, darker colours acting in a similar manner to black bodies, i.e. absorbing radiation. Therefore, dark-coloured eggs or other developmental stages have fewer problems in subzero conditions than light-coloured ones, with the added advantage that in the spring the dark-coloured stages will warm up faster and hatch or become active before the light-coloured ones. It is certainly noticeable that many Arctic insects show unusually dark coloration, e.g. bumble bees, fleas and butterflies (Downes 1962) and the majority of overwintering insect eggs are black, e.g. *R. padi*, whereas eggs laid in spring are green, e.g. *Panolis flammea* and *Pieris brassicae* (Lepidoptera: Pieridae). In addition, adults of the lacewings, *Chrysoperla carnea* and *C. plorabunda* (Neuroptera: Chrysopidae), are reddish-brown when in their winter diapause and light green in summer (MacLeod 1967; Hodek 1973; Tauber and Tauber 1973d). Apparently, there are advantages to be gained by having dark-coloured overwintering forms.

Intraspecific differences

Intraspecific differences are clearly shown in the two-spot ladybird, *Adalia bipunctata*, which overwinters as a adult. Brakefield (1985), working on populations of this insect in the Netherlands, was able to demonstrate

Table 5.8. *Changes in the frequency of the non-melanic and melanic forms of* Adalia bipunctata *in a hibernating cohort (after Brakefield 1985)*

Date	Number of non-melanics	Number of melanics		Total	% melanic
		Quad	Sexp		
2–3.12.81	1581	224	93	1898	16.7
26.1.82	362	89	42	493	26.6
1.4.82	237	47	36	320	25.94

Quad, quadrimaculata; Sexp, sexpustulata.

that the melanic forms were at an advantage over the non-melanic forms, particularly during the coldest part of the winter. This difference was very marked and the relative fitness of the non-melanic forms was calculated to be 0.55 (melanics = 1) (Table 5.8). It should be noted that, as with the other species discussed so far, overwintering mortality was very high – 83 per cent – but that suffered by the melanic forms was only 76 per cent as opposed to the 84 per cent suffered by the non-melanic forms. Colour was found to be the only difference – overwintering site selection was the same in both forms, as was the response to starvation. Thus the difference in winter survival was brought about solely by the advantage conferred by the dark pigment.

A similar situation is seen in the grain aphid *Sitobion avenae*. This overwinters both as eggs and as active parthenogenetic forms. The eggs are typically black and hatch in the spring to produce green aphids. During the spring and summer months, these aphids reproduce and give rise to both green and brown forms. There is some evidence to suggest that the brown forms are at an advantage during the summer due to increased reproduction brought about by the black body effect (Chroston 1982). In addition, although green clones can produce brown clones, brown clones cannot produce green clones. The majority of those individuals surviving the winter are brown, indicating that, as in the case of *A. bipunctata*, the darker forms are at an advantage over the lighter forms during winter.

Predation and colour

A dark coloration is also an advantage to those insects that overwinter as immobile subterranean or concealed forms. The pupae of many overwintering larvae, e.g. *Panolis flammea* and *Bupalus piniaria* (L.)

(Lepidoptera: Geometridae), although initially pale green, soon darken to dark brown and as they overwinter in the soil, become difficult to locate. As they are unable to avoid predation physically, being relatively immobile, this cryptic coloration affords some protection from those predators dependent on vision, e.g. birds. Overwintering aphid eggs laid in crevices of the bark, e.g. the sycamore aphid, *Drepanosiphum plata-noidis*, or in bud axils, e.g. *Rhopalosiphum padi*, are also harder to see once they have darkened from their initial green to black.

Conclusions

It is apparent that overwintering is a costly process and, in comparison to the summer stages, is a disadvantageous stage in the life cycle. However, tropical insects have often to enter diapause at the height of the dry season to escape conditions equally inhospitable as those experienced during winter by insects in temperate climates (Tauber and Tauber 1973d). There is thus no great advantage to be gained from living in the tropical parts of the world over temperate regions because, despite the opportunities of accelerated reproduction due to high temperatures afforded by the tropical habitat, a specialised stage in the life cycle is still required. In fact, some temperate insects, e.g. the sycamore aphid, *Drepanosiphum platanoidis*, also have a summer resting stage (Dixon 1966). Among temperate insects, although the cost of overwintering as an immobile specialised stage is expensive in terms of mortality and metabolic expenditure, there is at least a guarantee of some survival in all years, whereas winter active stages are at a greater risk in some years. A specialised overwintering stage is thus more of an asset than a liability.

6

Prediction and control

Introduction

In this chapter we draw together the many aspects of overwintering – survival of different overwintering stages, choice and success of overwintering sites, factors initiating the overwintering response – and discuss their importance to applied biology. In other words, what is the relevance of overwintering to the producer of a plant product, be it timber, cereals, fruit, ornamental plants or whatever? We will first illustrate the importance of a full knowledge of the overwintering habits of insects in forecasting the need for control measures to be taken, i.e. prediction, by reference to specific case studies where this method of prediction is actually employed, then discuss cases where the potential of such systems is under investigation or trial, and finally point out some examples where we feel that this method of prediction could be usefully and profitably utilised.

Ideally, a forecasting system should not involve any winter population sampling. After all, the weather during winter, even in temperate climates, is far from conducive to easy sampling. The ideal system should be based on the assessment of mortality from climatic measurements and from that the likelihood of outbreaks should be determined. One characteristic of all the systems described in this chapter is that although the overwintering stages are sampled, the sampling is done before the onset of winter or, if during the winter, only on stages easily accessible to the sampler. Predictions are then made on the basis of climate and or mortality, and forecasts made from this information.

In addition we will discuss the possibility of insect control by measures applied during the overwintering stage by use of published examples and predictive modelling. To prevent this from becoming a mere catalogue, we will describe the pest and its impact as well as the prediction system. In addition, we will, where possible, assess the value and accuracy of the forecasts provided. The examples that follow are not intended to be a comprehensive list of forecasting systems based on the overwintering stage, but were selected to illustrate particular techniques that have been developed or that could be developed.

Systems in use

The black bean aphid, Aphis fabae, *in southern England*

This is an outstanding example of the accurate forecasts that can be made by monitoring the overwintering stages in insects. The field bean, *Vicia faba*, is an important crop in southern England and spring-sown crops often suffer from infestations of the black bean aphid, *Aphis fabae* Scopoli. Damage effects can be severe and losses in yield of up to 46 per cent have been reported due to feeding by the aphids in addition to that caused by virus infection (Way and Heathcote 1966).

It was noticed during the 1960s that the numbers of eggs of *A. fabae* laid on its winter host (*Euonymus europaeus*) could be used to forecast the subsequent infestations on beans and sugar beet (Way and Banks 1968; Jones and Dunning 1972). In 1968, the Agricultural Development and Advisory Service (ADAS) in southern England, together with ento-mologists at Imperial College (University of London), initiated a fore-casting scheme based on monitoring winter egg populations and peak spring populations of *A. fabae* on *E. europaeus* bushes (Way *et al.* 1977). The spindle bush, *E. europaeus*, is widely distributed through southern and central England and 18 separate forecasting areas were designated. The eggs of *A. fabae* are laid in cracks on the bark of older wood and in the bud and twig axils of first year twigs. These are considered to be a good estimate of the overall bush population (Cammell *et al.* 1978).

Ten to twenty first-year twigs are taken at random from each bush within more than 300 designated indicator sites (Fig. 6.1). In addition, peak spring populations of *A. fabae* on spindle are estimated by further sampling in May. The egg counts are expressed as eggs per 100 buds and the peak aphid population as aphids/new shoot (Table 6.1). Five cate-gories are recognised, of which the three highest indicate a need for chemical control. Although this has proven a very successful forecasting scheme, attempts have been made in recent years to improve the accuracy of the forecast by linking autumn and spring catches from the Rothamsted

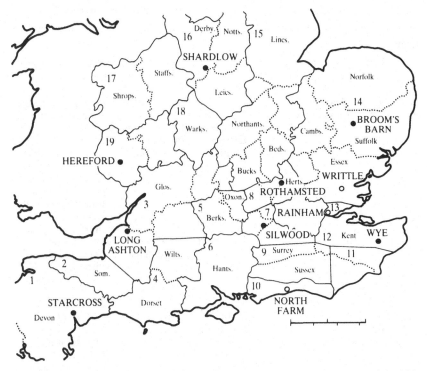

Fig. 6.1. *Aphis fabae* egg sampling areas and corresponding suction trap sites. (●) suction traps used in present evaluation (○) suction traps currently operating but data insufficient (redrawn after Way *et al.* 1981).

Insect Survey suction trapping with the egg counts (Way *et al.* 1981). A further refinement to the system has been the incorporation of information concerning the distribution and local abundance of the spindle trees in a particular forecasting area (Way and Cammell 1982). The forecasting scheme has an accuracy of over 90 per cent and is thus extremely valuable.

The bird cherry-oat aphid, Rhopalosiphum padi, *in Scandinavia*

The basic life cycle of this aphid has been described earlier (see p. 71). The summer stages of this aphid are severe pests of cereals in Scandinavia and northern Europe (Rautapää 1976; Leather and Lehti 1982) but they are only pests in Great Britain due to their ability to transmit barley yellow dwarf virus. The overwintering eggs are laid in the axils of the buds on young twigs of the bird cherry, *Prunus padus*, and hatch in spring. As has been stated earlier, the eggs of this aphid show a constant mortality

Table 6.1. *Categorization of egg and peak aphid numbers on spindle and the corresponding forecast and action necessary on field beans (after Way et al. 1977)*

Spindle			Beans	
Mean eggs/ 100 buds*	Mean peak aphids/new shoot	Category	Forecast	Meaning of forecast and action required
<1.1	<0.6	Extremely light	Unlikely damage	Less than 5% of plants likely to be infested initially. Not expected to cause economic injury. Chemical treatment unnecessary.
1.1–5.0	0.6–2.5	Very light	Possible damage	5–10% of plants likely to be infested initially. Chemical treatment necessary on some fields
5.1–25.0 25.1–100.0 >100.0	2.6–12.5 12.6–25.0 >25.0	Light Moderate Heavy	Probable damage	>10% of plants likely to be infested initially. Chemical treatment necessary on nearly all fields.

* Corrected for spindle abundance in Hampshire and East Anglia.

rate, which is correlated with the duration of the winter period (see Fig. 5.3). The first attempt at using egg counts as a forecasting system for *R. padi* was in England by Rogerson (1947), who tried to correlate egg numbers with aphid populations on oats. This was not a success due to the short duration of the work and the low number of trees sampled. However, during the late 1970s and early 1980s workers in Sweden and Finland began to study the possibilities of using winter egg counts as a forecasting system (Johansson 1981; Leather and Lehti 1981), and since 1988 this system has been used in Finland to advise farmers of the likelihood of control measures being required (Kurppa 1989).

Prunus padus trees are sampled at 50 to 110 sites throughout southern and central Finland during the winter. Twenty young branches (0.5–1.0 m long) are removed from each of five healthy trees or bushes in a 1 km vicinity and then forwarded to a central laboratory where the number

Table 6.2. *Egg levels at which aphid control measures are likely to be required for* Rhopalosiphum padi *in Scandinavia (after Leather 1983)*

Autumn counts (eggs/100 buds)	Spring counts (eggs/100 buds)	Category	Forecast	Action required
>75	>40	Heavy	Probable damage	Tiller counts required in June and if more than two aphids/ tiller chemical control necessary
50–74	23–39	Moderate	Possible damage	Tiller counts required in June
30–49	15–24	Light	Slight risk of damage	If warm spring, tiller counts required in June
<30	<15	Very light	Unlikely damage	None

of eggs per bud is counted. Work has also shown in both Finland and England that the number of eggs present is correlated very well with the peak population achieved by this aphid on *P. padus* during spring (Leather 1983; Kurppa 1989; Leather unpublished data). A single count in either autumn or in spring prior to egg hatch would be sufficient due to the predictable nature of egg mortality, but employment of both counts would improve the accuracy of this prediction. Although, as in the case of *A. fabae*, other factors, e.g. weather, are likely to affect *R. padi* population build-up, it has been possible to produce a table of threshold values for use in forecasting the likelihood of damage (Table 6.2). Another advantage of this system is that is possible to produce a map of a country, region by region and give general indications to farmers very early in the winter. This was first done in Finland during 1981 (Fig. 6.2; Leather 1982) and since then has been used yearly with great success (Fig. 6.3; Kurppa 1986). This system has an advantage over existing systems based on summer populations, e.g. Wiktelius (1982), in that many months notice can be given to the farmers, rather than a week or two as is the case at present. Although this forecasting system has not yet been tested fully it is likely to overestimate the need for control rather than underestimate. It will also only be suitable for use in countries such as Finland and

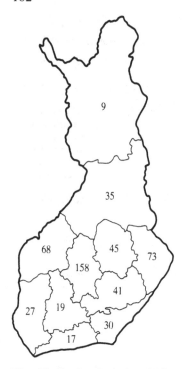

Fig. 6.2. Regional results of *Rhopalosiphum padi* egg sampling in Finland, autumn 1981 (after Leather 1982).

Sweden where the only method of overwintering is the egg stage. In countries such as Great Britain, where both eggs and active stages overwinter, a different system would have to be used. This is discussed later.

Potato aphids in Britain

The production of virus-free seed potatoes is an important aspect of agriculture in the north and east of Scotland. Several studies have implicated a range of aphid species as efficient vectors of potato viruses (e.g. Tamada and Harrison 1981) amongst which the most important is *Myzus persicae*, the peach-potato aphid (Van Harten 1983: Harrington and Cheng 1984).

The large numbers of records of *M. persicae* successfully overwintering on weeds in Scotland (Fisken 1959) and the scarcity of its woody primary host plants, indicate that the species overwinters mainly in the anholo-

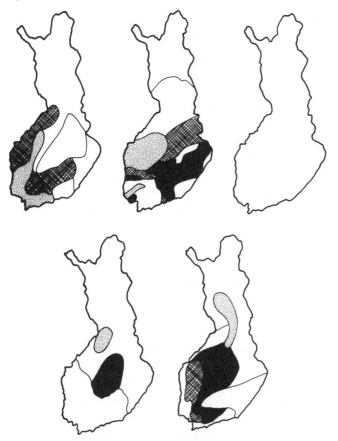

Fig. 6.3. The density of overwintering eggs of *Rhopalosiphum padi* on *Prunus padus* in different areas of Finland 1981–5 ■ >40 eggs/100 buds, ▧ 25–39 eggs/100 buds ▨ 15–24 eggs/100 buds, □ <15 eggs/100 buds) (after Kurppa 1986).

cyclic state in the seed potato growing areas. The relatively severe winters experienced in most years in these regions limit the success of over-wintering of these aphids and, consequently, their spring migration to potato crops. For example, between 1970 and 1976, a series of mild winters resulted in early and large spring migration of *M. persicae* and this increased the incidence of potato leaf roll virus (Howell 1977). Subsequent cold winters returned the virus incidence to low levels, but the need for forecasting high risk years was recognised. As aphids flying early in the crop growing season have a greater effect on the spread of aphid-borne viruses than those flying later in the year (Cadman and Chambers 1960), the size and timing of aphid migrations from winter hosts to potatoes in

Table 6.3. *Calculated values of (a), (b), (r) and the standard error of (b)*
in the regression, potato aphid arrival date in the East Craigs 12.2 m
suction trap against difference in accumulated day-degrees above and
below 5 °C (after Turl 1980)

Species	Period showing highest significance	(a)	(b) ±SE	(r)
Myzus persicae	February	19.1	−0.51 ± 0.09	−0.912
Macrosiphum euphorbiae	February	13.7	−0.50 ± 0.07*	−0.947
Aulocorthum solani	January–April	45.6	−0.31 ± 0.05*	−0.942
	February	18.3	−0.63 ± 0.17*	−0.832

* SE within 95% limit for *t*; a, coefficient constant; b, slope; r, regression coefficient.

spring are two important factors in determining the risk from viruses in
any year.

The timing of the spring migration was investigated by Turl (1980).
Using data obtained from the Rothamsted Insect Survey, 12.2 m suction
traps and records of winter temperatures, she found negative correlations
between the date of the first catch and the number of day degrees above
and below 5 °C, for several species of aphids, including *M. persicae*. The
strength of these relationships varied between species but was generally
very high, accounting for between 80 and 95 per cent of the variance in
time of the first catch (Table 6.3). A similar relationship has since been
described for *M. persicae* caught in suction traps sited in England (Bale
et al. 1988). This potential for forecasting the time of the spring migration
of potato aphids now needs further testing and verification, although the
system is now in general use in England.

The investigation of techniques for forecasting the size of the early
season aphid flight concentrated initially on a study of overwintering
populations. The survival of anholocyclic *M. persicae* populations during
winter has been found to be related to temperature in at least two field
studies (Harrington and Cheng 1984; Walters 1987), with poorer survival
being recorded in colder winters (see Chapter 2). In Scotland the
population dynamics of *M. persicae* overwintering on cabbages showed
a similar pattern in each year of a 6-year field experiment. Population
sizes increased in late autumn and early winter, usually reaching a peak
in early December. At the start of the coldest winter weather, in January
and February, populations became progressively smaller until reaching
their lowest point in March, when they again started to increase until
spring and early summer, when they gave rise to alatae. These alatae flew

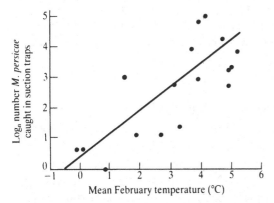

Fig. 6.4. The relationship between mean February temperature and the number of *Myzus persicae* caught in suction traps during the period when potatoes are at risk from infection by aphid-borne viruses.

to the summer hosts, amongst which are potatoes. The reduction in the size of the populations between December and March was significantly related to the mean daily temperature to which they were exposed (Walters 1987). As the size of the spring migration of aphids is related to the size of the population on their overwintering hosts in spring (Leather and Walters 1984), the relationship between winter temperatures and the size of the aphid flight was investigated. Using data from 17 years of trapping using 12.2 m suction traps, a strong correlation was demonstrated between the natural logarithm of the numbers of aphids caught during the period when potato crops are most susceptible to damage by viruses and winter temperatures from the coldest months of the year (Fig. 6.4). This relationship accounted for 75 per cent of the variance in the size of the aphid flight (Walters 1987) and provides a basis for forecasting the size of the migration, which again requires further testing and verification.

Thus both the size and timing of the flight of aphid virus vectors and potato crops may be forecast using this technique. Such forecasts will permit prophylactic measures, such as application of insecticidal granules at planting to be used more effectively. However, as forecasts of aphid flights based on winter temperatures are most accurate in regions where aphids overwinter anholocyclically they should be made on a regional rather than on a national scale (Walters and Dewar 1986). The proportion of the aphids caught that are actually carrying virus particles will be determined in part by the amount of virus present in the environment. This background level of virus is known to vary between years and can be estimated in a particular season using an assessment of the virus levels in potato crops from the previous year. The development of tests to determine if individual aphids are carrying virus particles may achieve the same object more readily.

The pine beauty moth, Panolis flammea, *in Scotland*

The pine beauty moth, *Panolis flammea*, is an indigenous moth species occurring on Scots pine, *Pinus sylvestris*, throughout Britain. In the rest of Europe it is a severe pest of this species and has been reported as such for almost 200 years (Klimetzek 1972). A simple forecasting scheme based on the number of overwintering pupae present in the soil was developed in Germany during the late 1920s (Schwerdtfeger 1934). A critical damage threshold was calculated (75 per cent defoliation) and a figure of 1 pupa/m^2 determined to be the population level at which control measures should be taken. In Britain there has never been an outbreak of *P. flammea* on *P. sylvestris*. However, during the 1970s numbers of *P. flammea* built up in plantations of introduced lodgepole pine, *Pinus contorta*, culminating in a series of outbreaks in northern Scotland during 1976, 1977 and 1978, in which large areas of *P. contorta* were completely defoliated and hence killed (Stoakley 1979). A forecasting scheme based on that used in central Europe was therefore designed to insure against the possibility of a recurrence of this problem. Further outbreaks have since followed but these have been successfully controlled with only minimum tree loss due to the success of the prediction system discussed below (Watt and Leather 1988).

Panolis flammea feeds as a larva on pine foliage during the summer months and leaves the trees during August to pupate in the needle litter layers on the forest floor. Sampling is usually carried out during autumn before snow falls and sampling becomes impossible. It has been determined that trees do not become at risk until they are 12 to 15 years old so only stands of this age are designated as 'at risk' and surveyed. Unlike the European situation on Scots pine, where control is considered necessary at pupal levels of 1/m^2, the density of pupae at which control measures become necessary in Scotland is 15/m^2. However, further sampling takes place in these cases at the egg stage. If the number of eggs exceeds 1000/tree then control measures are indicated. There is a good relationship between pupal numbers and egg numbers (Fig. 6.5) but spring weather conditions can affect this considerably, particularly at pupal levels close to, or at, the critical threshold value. A model has been developed to

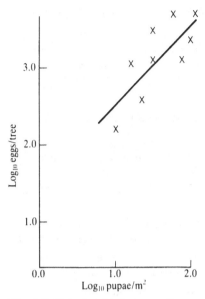

Fig. 6.5. Relationship between the numbers of eggs of *Panolis flammea* (log n + 1) and number of pupae/m^2 (log n + 1).

predict the effect of weather and will be incorporated into the standard sampling scheme (Leather *et al.* 1985). Local topography and the extent of previously defoliated areas also need to be taken into account. To date this forecasting system (where evidence has arisen due to failure of control programmes or failure to take control measures) has an accuracy of over 90 per cent, and can thus be regarded as highly reliable.

The pine looper moth, Bupalus piniaria, *in Britain*

The pine looper moth, *Bupalus piniaria* (L.) (Lepidoptera: Geometridae), is one of the most serious defoliating pests of pine in central and eastern Europe, Scandinavia and CIS. It is indigenous to Great Britain and is common in most pinewoods. The first outbreak in Britain occurred on Scots pine in the English Midlands in 1953; others have since occurred at intervals throughout northern Britain. The overwintering stage is the pupa and, as in *P. flammea*, pupation occurs in the soil beneath the trees. Accurate counts can be made by counting the number of pupae found in the top 15 mm of soil in circular plots 0.25 m^2 in area. Ten plots are spaced equidistantly on a straight transect through each 10 hectare area sampled. A body of data based on these pupal counts has been built up

Table 6.4. *Categorization of pupal counts and corresponding forecast and action necessary for* Bupalus piniaria *on Scots pine in Britain*

Mean pupae/m^2	Category	Forecast	Action required
0–10	Normal	Unlikely damage	None
11–30	Suboutbreak level	Slight feeding damage	Visual check for localized damage
> 30	Outbreak level	Moderate to severe damage	Egg counts in summer: insecticidal control if necessary

by the Forestry Commission over a period of over 30 years (Bevan and Brown 1978; Barbour 1985) and a scale of numbers has been developed as a predictor of possible damage (Table 6.4). Counts of more than 10 pupae/m^2 are regarded as critical, and require further action to determine damage done, egg density and the need for insecticidal control.

This forecasting scheme has never underpredicted or failed to predict outbreaks before they occur, although it has predicted outbreaks that have not occurred. This is a result of the uncertain nature of the effect of the natural enemies of *B. piniaria* and the interaction of both their populations and that of their hosts with weather (Barbour 1985).

The Douglas fir tussock moth, Orgyia pseudotsugata, *in North America*

The Douglas fir tussock moth, *Orgyia pseudotsugata* (McD) (Lepidoptera: Lymantriidae), is a major defoliator of Douglas fir (*Pseudotsuga menziesii*) and true firs (*Abies* spp.) in the western United States and Canada. Outbreaks of this insect occur periodically in North America and have been a severe problem since 1953 (Wickman *et al.* 1981; Mason *et al.* 1983). The overwintering stage is the egg, which is laid in large masses by the wingless female on the foliage of the tree. Egg-hatch occurs in the late spring or early summer. Egg-mass sampling is used to monitor and evaluate populations in order to predict population trends and plan potential control measures. In addition, the timing of egg-hatch is used to predict the scheduling of spray operations. The number of eggs laid per egg-mass is not a constant and is affected by geographical location and phase of the outbreak (Mason *et al.* 1977) as well as the environmental and nutritional factors affecting the parent (Beckwith 1976). A method was thus devised to relate egg-mass weight with the number of eggs

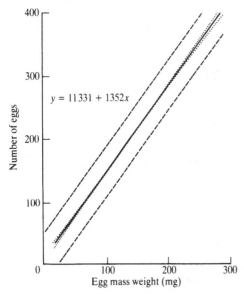

Fig. 6.6. Common curve for estimating the number of eggs of Douglas fir tussock moth (*Orgyia pseudotsugata*) from egg-mass weight. —— common curve, ·····
95% confidence interval of mean, ――― 95% confidence interval of individual (after Beckwith *et al.* 1978).

present, which is applicable to any site or phase of the outbreak (Fig. 6.6). Critical values for egg counts were derived from studies of parasitism rates and foliage consumption rates by the larvae and have been used successfully in outbreak prognosis models (Mason and Wickman 1988).

Grasshoppers in Canada

Grasshopper outbreaks on cereals have been commonplace in Saskatchewan since the turn of the century and were accepted somewhat fatalistically by the farming community. However, since 1931 populations of adult grasshoppers and eggs (Orthoptera: Acrididae) have been counted in all arable land areas in Saskatchewan and rated in four categories of potential outbreak hazard. These are then mapped for their respective locations and a forecast map produced indicating areas of probable light, moderate, severe or very severe outbreak the following year (Riegert 1967, 1968). An example of maps for 2 years is shown in Figure 6.7. These maps have been accepted by agriculturalists as a standard reference and used for crop protection decision-making throughout the province.

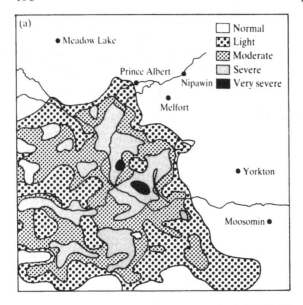

Fig. 6.7. (a) and (b) Hazard forecasting for grasshopper outbreaks in Saskatchewan, Canada (after Riegert 1967).

Over the years the first simple estimations done solely by number of eggs have been refined and now soil type and oviposition site are also taken into account before a hazard rating is given (Table 6.5).

Table 6.5. *A table of ratings used in Saskatchewan since 1948 to predict grasshopper outbreaks from overwinter egg counts (after Riegert 1967)*

Category	Rating	Eggs/m²		Roadside pods
		Field		
		Pods	Eggs	
Normal	1	0.08	0.9–1.5	Up to 0.3
Light	2 –	0.11	1.8–2.1	0.6–0.9
	2	0.15	2.4–3.0	1.2–1.8
Moderate	3 –	0.23	3.4–4.6	2.1–3.0
	3	0.30	4.9–6.1	3.4–4.6
Severe	4 –	0.46	6.4–9.1	4.9–6.1
	4	0.60	9.4–12.2	6.4–9.1
Very severe	5 –	0.91	12.5–18.3	9.4–12.2
	5	>1.2	>18.5	>12.4

Width (m) of field margin, ditch edge etc	% population used
0.3–1	25
1.2–1.8	50
2.1–2.7	75
over 3.0	100

Extent and frequency of egg masses	% population used
Very small spots 0.3–0.7 m²	10
Small, beds 0.7–1.8 m diameter	25
Medium, beds 1.8–3.0 m diameter	50
Large, beds over 3.0 m diameter	75 +

Leatherjackets, Tipula *spp. in Britain*

Leatherjackets, the larvae of crane flies (Diptera: Tipulidae), are native to grasslands and are very important pests of pastureland, hay crops and crops that are grown in land that has been recently ploughed from grassland. *Tipula paludosa*, the most common species in Britain, lays its eggs in the soil during the autumn. The larvae hatch soon after and feed on the roots of plants throughout the winter until the following summer, causing severe damage when populations are high. Pupation then takes place, and adults emerge in the early autumn (Ministry of Agriculture, Fisheries and Food 1979).

Surveys for these insects are carried out in several regions of England and Wales by the Agricultural Development and Advisory Service. Soil core samples are taken from permanent pasture sites during the winter. These provide the basis for the forecasts of likely damage to crops the following spring. For example, in cereals the economic threshold is 0.5 million larvae/hectare. Crops at risk are then checked regularly and further sampling is carried out if feeding damage is suspected. Control measures will then be taken if justified (Ministry of Agriculture, Fisheries and Food 1979).

The wheat bulb fly, Delia coarctata, *in Britain*

The wheat bulb fly, *Delia coarctata* (Fall.) (Diptera: Anthomyiidae), is a severe pest of winter wheat in Great Britain. The eggs are laid during July and August onto bare soil or in soil under a root crop, just beneath or on the surface. The eggs hatch during the winter (January) and the young larvae bore into the central shoots and tillers (Ministry of Agriculture, Fisheries and Food 1982).

As with leatherjackets, the Agriculture Development and Advisory Service monitor for this pest. Egg populations in fallow and partially fallow fields are sampled during August–September, with a threshold value of 2.5 million eggs/hectare being used (Oakley and Uncles 1977). As this is a presowing survey, a number of control strategies are open to use and the forecast is extremely useful.

Systems in development

In this section we concentrate on forecasting systems based on the overwintering stages of insects that are still in the developmental stage. Furthermore, in some cases we suggest the use of the overwintering stage as a population predictor where it has not so far been used in this manner.

In many cases knowledge of overwintering success is only part of the information that is required to develop the forecasting scheme, as several other factors will need to be quantified before a fully reliable picture of the risk from virus infection in any one year can be constructed.

Cereal aphids in Scotland

Many of the cereal-infesting aphids in Scotland have the potential to overwinter either completely or partially anholocyclically. This makes

Table 6.6. *Calculated values of (a), (b), (r) and the standard error of (b) in the regression, cereal aphid arrival in the East Craigs 12.2 m suction trap against differences in accumulated day-degrees above and below 5 °C (after Turl 1980)*

Species	Period showing highest significance	(a)	(b) ±SE	(r)
Sitobion avenae	February	22.1	−0.42 ± 0.10	−0.868
Sitobion fragariae	February	30.0	−0.39 ± 0.12	−0.807
Metopolophium dirhodum	November–December	33.5	−0.16 ± 0.03*	−0.919
	April	56.4	−0.39 ± 0.09*	−0.878
Metopolophium festucae	February–March	17.9	−0.29 ± 0.05*	−0.909
	February	13.4	−0.45 ± 0.10*	−0.884
Rhopalosiphum padi	None			
Rhopalosiphum insertum	February–April	52.0	−0.42 ± 0.19	−0.884

* SE within 95% limit for *t*; a, coefficient constant; b, slope; r, regression coefficient.

forecasting their first appearance and abundance on crops difficult in comparison with species like *A. fabae* in England or *R. padi* in Scandinavia, which only overwinter as eggs. Even in areas where wholly holocyclic overwintering can be expected, the situation is not clear-cut. Shands *et al.* (1961) attempted to forecast the abundance of potato aphids in Maine (USA) but found that other non-pest aphids were also laying eggs on the same hosts. Turl (1980, 1983) has been concerned with aphids infesting cereal crops in Scotland both in relation to the direct feeding damage they inflict on the crop and their potential as virus vectors, in particular *Sitobion avenae*, *Rhopalosiphum padi* and *Metopolophium dirhodum* which are vectors of barley yellow dwarf virus, a severe disease of cereals worldwide.

Using suction trap data to ascertain the time of flight to cereal crops in spring and day-degrees above and below 5 °C Turl has been able to show a strong negative correlation between winter temperatures and date of arrival for six different species of cereal feeding aphid, which explained between 80 and 95 per cent of the variance, depending on species (Table 6.6). Such strong relationships have great potential as a forecasting technique. The ultimate forecasting system would depend very much on the crop – spring-sown barley, for example, is most at risk when early aphid arrival coincides with above average June temperatures (Sparrow 1974). More work has to be carried out linking arrival date with subsequent population development and yield loss before this system can be considered a true forecasting scheme.

Table 6.7. *Winter weather,* Prunus padus *abundance and peak*
Rhopalosiphum padi *catches in Britain*

Deviation from mean winter weather (°C)	*P. padus* abundance category	Likelihood of BYDV infestation
+3	0, 1, 2 or 3	High
+2	0, 1, 2 or 3	High
+1	0, 1, 2 or 3	High
0	1, 2, 3	Moderate
−1	2, 3	Moderate
−2	2, 3	Slight
−3	3	Slight

BYDV, barley yellow dwarf virus; 0 = no *P. padus* present in a 100 km radius; 1, <3 10 km squares containing *P. padus* in a 100 km radius; <6 10 km squares containing *P. padus* in a 100 km radius; >6 10 km squares containing *P. padus* in a 100 km radius.

The bird cherry-oat aphid, Rhopalosiphum padi, *in Britain*

As was seen earlier, *Rhopalosiphum padi* overwinters both as the egg stage (holocyclically) and anholocyclically in Britain. Thus the use of winter egg counts alone does not give a good relationship between aphid numbers and eggs laid. In addition, the distribution of the winter host, *Prunus padus*, is not uniform, being almost totally absent from southern England (Leather 1980b). Thus, any forecasting system must take both these factors, and winter weather, into account. An early attempt at correlating winter weather, *P. padus* density and peak suction trap catches of *R. padi* for a limited number of sites indicated that correlations between aphid numbers and *P. padus* density only occurred after severe winters, i.e. when all or the majority of the anholocyclic forms had been killed (Leather 1980b). A more recent piece of work (Knight and Leather in preparation) has correlated all three successfully and can predict the likelihood of the occurrence of greater than average numbers of *R. padi* (Table 6.7). However, this approach requires considerable refinement before it can be used successfully.

The strawberry aphid, Chaetosiphon fragaefolii, *in Canada*

The most efficient vector of strawberry viruses in British Columbia is the strawberry aphid, *Chaetosiphon fragaefolii* (Cockerell). The eggs of this aphid are laid on the leaves of strawberry plants during October and November. Using a simple accumulated day-degree model (threshold value of 4.96 °C), Frazer and Raworth (1984) have been able to predict egg-hatch at least 3 weeks before it actually occurs. This allows the time of earliest flight of aphids to new plantings of strawberries to be predicted and the scheduling of transplanting and insecticide application can be completed well in advance.

The bertha armyworm, Mamestra configurata *in Canada*

The bertha armyworm, *Mamestra configurata* (Lepidoptera: Noctuidae) is a severe pest of canola, *Brassica napus*, and *B. campestris* in Canada (Turnock and Philip 1977). Damage is widespread but of a sporadic nature. The moth overwinters as a pupa in the soil, emerging to lay its eggs on the leaves of canola in the spring. Pupal number can easily be assessed during the autumn by soil sampling. Pupal survival can be estimated accurately from daily soil temperatures using a computer simulation model and by taking into account snow cover (Lamb *et al.* 1985). This system is in the early stages of development but the strong correlations between outbreak distribution and areas where winter soil temperatures are highest (Fig. 6.8), indicate that it could become a very useful forecasting system.

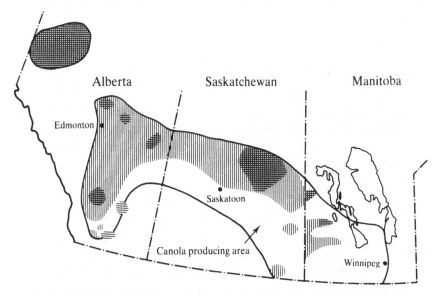

Fig. 6.8. The average of winter days with soil temperatures of −10 °C or less at 5 cm depth from 1969–83. The distribution of bertha armyworm outbreaks ⫽ 1970–3, □ 1979–83 and the area of canola production (solid outline) in use (after Lamb *et al.* 1985).

The European sawfly, Neodiprion sertifer, *in Britain*

The sawfly, *Neodiprion sertifer* (Hymenoptera: Diprionidae) is a chronic pest of young lodgepole pine crops in northern Britain (Stoakley 1989). Its eggs are laid in short rows in the needles of the current year's foliage and spend the winter there in a dormant stage. The young larvae hatch from the eggs and start feeding on the needles during May. Populations are assessed from the egg numbers. Counts are made along transects consisting of single rows of trees so as to cover 40 hectares of the block. Every tenth tree is sampled until ten trees in a transect have been sampled. The leader is removed from each tree and examined for egg clutches. This is repeated for each first-whorl, second-whorl and third-whorl branch in turn. The lower whorls have only one branch sampled. The data collected are then entered into a computer and the number of clutches per tree is produced. This system does not, as yet, include a winter survival component and follow-up counts of the larvae are made in spring at those sites where populations are deemed to be in the 'at risk' category. Information is available for winter survival of *N. sertifer* eggs in Canada (Sullivan 1965) and it has been suggested that prediction of overwintering survival should not present a great problem. If this component were added to the British system, the efficacy of the forecasting system would be greatly enhanced.

Systems still to be developed

There are many insects in which the overwintering stage would prove to be an easy and convenient predictor of the populations likely to be encountered during the following season. However, in many cases the necessary background research has only just begun. In this section we highlight a few of the many examples.

The green spruce aphid, Elatobium abietinum, *in Britain*

The green spruce aphid, *Elatobium abietinum* (Homoptera: Aphididae), is a severe pest of spruce in Britain (Carter and Nichols 1988) causing needle loss and discoloration of a number of spruce species. It has become a severe problem to Christmas tree growers and also causes incremental losses in mature trees (Carter 1977). In Britain, this aphid has no egg stage and overwinters as the parthenogenetic female stage. It is thus susceptible to the effects of winter weather. Outbreaks of the aphid on

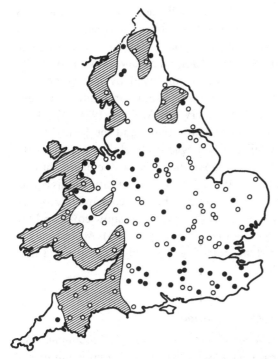

Fig. 6.9. Distribution of forests severely attacked by the green spruce aphid in spring 1971 and the occurrence of temperatures below $-8\,°C$ from December 1970 to March 1971. ▨ region of severe attack in spring 1971, ● stations recording air minima of $-8\,°C$ or below, December 1970–March 1971, ○ stations with no recorded air minima as low as $-8\,°C$, December 1970–March 1971 (redrawn after Carter 1971).

Sitka spruce, *Picea sitchensis*, have been correlated with warm winter weather in England (Fig. 6.9). This would appear to be a fairly simple system from which to develop a scheme that gives accurate forecasts of outbreaks.

The gypsy moth, Lymantria dispar, *in North America*

The gypsy moth, *Lymantria dispar* (L.) (Lepidoptera: Lymantriidae), has a worldwide distribution and is a severe pest of both forest and shade trees (Leonard 1974). It is widespread in North America, despite its relatively recent introduction (1868) and the large expenditure of money and time in attempts to prevent its spread (Leonard 1974). The eggs are laid in clusters on the host plants during early summer and remain in diapause during winter, hatching as bud-break begins the next spring (Montgomery and Wallner 1988). The larvae feed on the foliage and cause extensive defoliation to a number of tree species. At present detection of gypsy moth outbreaks is mostly based on aerial defoliation surveys with follow-up ground egg-mass counts. This could be improved by incorporating a predictive element into the system. Preventive spray programmes usually follow the egg-mass counts (Montgomery and Wallner 1988). Work on the cold-hardiness of winter moth eggs in Canada and the United States has shown that winter mortality of the eggs is closely correlated to the length of exposure to low winter temperatures and to the degree of cold experienced (Madrid and Stewart 1981). Further elaboration of this earlier model shows that events in the forest can now be predicted and interpreted using micrometeorological techniques to account for snow cover and the effects of sudden drops in temperature (Waggoner 1985). Accordingly, a very accurate forecasting system indicating the need to spray could now be developed for this moth.

The turnip moth, Agrotis segetum, *in Britain*

The turnip moth, *Agrotis segetum* (D & S) (Lepidoptera: Noctuidae), is a serious pest of vegetable crops, feeding on the stems of a large number of plants, such as potatoes, turnips and lettuce (Ministry of Agriculture, Fisheries and Food 1983). The overwintering stage is the larva, which spends the winter in the soil, pupating in April–May. The current monitoring system is based on the use of light and pheromone traps to monitor adult activity. Temperature and rainfall data are then used to provide an index of egg development and larval survival (Ministry of Agriculture, Fisheries and Food 1983). It seems to require only a small step to extend the predictive part of this forecast to take account of winter mortality and to be able to issue an even earlier warning of possible damage from soil sampling for larvae in the autumn.

The cereal leaf beetle, Oulema melanoplus, *in the United States*

The cereal leaf beetle, *Oulema melanoplus* (L.) (Coleoptera: Chrysomelidae), is a pest of cereals in the United States (Sawyer and Haynes 1985). Detailed studies of the effect of low temperatures on the mortality of adult beetles (Casagrande and Haynes 1976) and the effect of regional differences in cold tolerance have allowed a predictive model for the mortality of overwintering beetles to be constructed. However, further

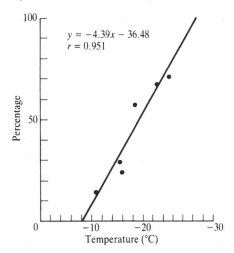

Fig. 6.10. Overwintering mortality of eggs of *Rhyacionia buoliana* in Germany (after Bogenschutz 1976).

development is required before this system is likely to be of more than limited use in forecasting outbreaks.

Other insects

These approaches may be useful for many other insect species. The pine shoot moth, *Rhyacionia buoliana* (D & S) (Lepidoptera: Tortricidae), is a defoliating pest of pine in both North America and Europe. It overwinters as an egg and strong correlations between overwintering mortality and temperature have been found (Pointing 1963; Bogenschutz 1976) (Fig. 6.10). The codling moth, *Cydia pomonella* (L.) (Lepidoptera: Tortricidae), a severe pest of orchard trees in Poland and elsewhere (Labanowski 1981) as an example where this approach is already in use. Other insects for which scope for similar work exists are the balsam woody aphid, *Adelges piceae* (Ratz.) (Homoptera: Adelgidae), in Canada (Greenbank 1970); the spirea aphid, *Aphis citricola* van der Goot (Homoptera: Aphididae), on citrus in Japan (Komazaki 1983); the boll weevil, *Anthonomus grandis* Boleman (Coleoptera: Curculionidae), in the United States (Price *et al.* 1985; Fuchs and England 1989) and the tent caterpillar, *Malacosoma disstria* Hubner (Lepidoptera: Lasiocampidae), again in the United States (Wetzel *et al.* 1973). Many similar examples could be cited but a catalogue of these would be of little value. Suffice to say, that where an overwintering stage exists that can be easily sampled, then a potential predictive system exists for that insect species.

Control

Control of an insect pest at the overwintering stage can seem quite an attractive proposition. After all, the pest species is usually at its lowest population level of the year and, as the preceding chapters have shown, its individual members are usually inactive and unable to escape predators or control operators. However, a major problem that is often encountered is that the devices used to escape the rigours of winter can also be very effective means of escaping the control measures. The chemicals needed to overcome insect pests in the winter usually have to be quite powerful – tar washes, for example, are used in orchards for control against a number of insect and mite species that overwinter as exposed eggs. In fact tar oils and allied chemicals can be used very successfully in this context to control insects such as aphids, winter moth, *Operophtera brumata*, tortrix moths and scale insects on the dormant stages of fruit crops (Ivens 1988). These cases are characterised by the pest being *in situ* on the crop that they are going to attack in the spring and by the fact that the crops are very localised.

Overwintering can also be a problem to the pest control agent. This is especially true in the case of biological control. Diapause affects the seasonal synchrony between pests and their natural enemies and the susceptibility of the pests to their natural enemies. It can also interfere in the rearing of natural enemies for use as biological control agents and has complicated the use of natural enemies from different parts of the world (Tauber *et al.* 1986).

In this section we attempt to answer two questions: (i) how amenable is the overwintering stage to control; and (ii) how relevant is control at the overwintering stage to present day control strategies?

The direct approach

Control at the overwintering stage, as has been stated earlier can appear to be a potentially useful approach. The bird cherry-oat aphid, *Rhopalosiphum padi* (L.), which is a serious pest of cereals in Scandinavia (Rautapää 1976), overwinters on bird cherry, *Prunus padus*, trees which are in easily locatable sites (Leather and Lehti 1981). It has been suggested that application of an ovicide to trees in areas where egg populations are high would be an effective and cheap way of reducing the pest status of this insect (Leather and Lehti 1982). Another approach would be to reduce the numbers of bird cherry in farming areas, as their abundance is a major factor in determining aphid numbers (Wiktelius 1988; Leather *et al.* 1989). This would be the cheapest solution in the long run, as not all the trees would have to be removed because intraspecific competition for egg-laying sites would keep the overwintering aphid population in the surviving trees to a manageable level (Leather 1990). Unfortunately, the trees are a much-loved feature of the Scandinavian landscape, and not all the land with trees is owned by the farming community. Government intervention would be required to implement tree destruction, but nevertheless one of these two approaches would be environmentally and economically more effective than the current method of control in which pesticides are applied during the growing season.

Host plant manipulations

Another approach to controlling pest species through the overwintering stage is by way of plant breeding (Eidt and Little 1968). Most phytophagous insects time their development to that of their hosts, particularly in areas where plant growth is very seasonal. Any way in which the synchronisation of host and pest can be broken is potentially harmful to the insect. This would be of most importance in spring-feeding insects, where the degree of synchrony of insect emergence and bud-break is of paramount importance. For example, *Quercus rubra* L. trees that flush late escape attack by the winter moth, *Operophtera brumata*, (Embree 1967) in the early summer. The pine beauty moth, *Panolis flammea* (D & S), a pest of pine in Europe, is also very dependent on the synchrony of bud-burst and egg-hatch (Watt 1987). It has been shown that the use of plant growth inhibitors is effective in delaying bud-burst in balsam fir, *Abies balsamea* (L), and white spruce, *Picea glauca* (Moench), and that this could prove an effective control against the spruce budworm, *Choristoneura fumiferana* (Clemens) (Lepidoptera: Tortricidae), which is the most destructive pest of coniferous forests in North America (Eidt and Little 1968). Much work still needs to be done.

Insect control methods based on host manipulation have received inadequate attention and, despite the fact that both Gerhold (1966) and Hanover (1975) have called for an increased interaction between entomologists and plant breeders, little has been achieved in this direction over the past two decades.

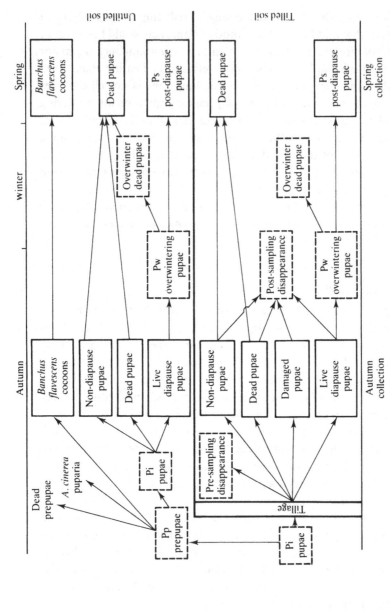

Fig. 6.11. Flow diagram of *Mamesta configurata* from the prepupal stage in autumn to the postdiapause pupal stage in spring. Variables enclosed in solid line were calculated from autumn and spring collections. Pp, prepupae; Pi, initial number of pupae formed; Pw, number of living pupae at the beginning of winter; Ps, number of living postdiapause pupae in the spring (after Turnock and Bilodeau 1984).

Cultural practice

Another approach to the control of an insect pest through its overwintering stage is to change or alter the winter habitat of the insect in question so that it is no longer suitable. With agricultural pests this can be done fairly simply by changes in tilling procedures.

The bertha armyworm, Mamestra configurata, *in North America*

The bertha armyworm, *Mamestra configurata*, a pest of canola, overwinters as a pupa in the soil. In untilled soil the survival rate of the pupae over the winter is twice that of the pupae in tilled fields (Turnock and Bilodeau 1984). Mortality in untilled sites was due primarily to the effects of parasitic wasps, in particular *Banchus flavescens* Cress. (Hymenoptera: Ichneumonidae), whereas that in tilled fields was due to the actions of predators and increased winter mortality due to reduced snow cover. On untilled sites the remaining stubble allowed snow to accumulate more than on tilled sites (Turnock and Bilodeau 1984). Tillage also resulted in mortality of pupae from direct damage. In this case cultural practices increased the number of possible mortality effects (Fig. 6.11).

The European corn borer, Ostrinia nubilalis, *in North America*

Tillage treatments are also effective means of control against the European corn borer, *Ostrinia nubilalis* (Hubner) (Lepidoptera: Pyralidae), in the United States. This insect overwinters as a fifth instar larva on the corn stubble, cobs, weed hosts and other plant material. Any form of ploughing has a significant effect on the overwintering survival, and adult emergence of *O. nubilalis* can be reduced by up to 98 per cent and on average 70 per cent (Umeozor *et al.* 1985).

The asparagus aphid, Brachycolus asparagi, *in the United States*

The asparagus aphid, *Brachycolus asparagi* Mordvilko (Homoptera: Aphididae), is a recent but severe pest of asparagus in the United States (Halfhill *et al.* 1984). It overwinters as the egg stage on the foliage of its host plant, hatching in spring to feed on the leaves. A combination of cultural treatments involving the use of burning, mowing and foliage removal in the autumn resulted in marked reductions in aphid populations in the spring (Table 6.8). Both absolute numbers and infestation rates were significantly reduced by autumn and spring tillage treatments. This is a clear demonstration of where an inexpensive change in cultural methods can result in a significant economic gain.

Table 6.8. *The effect of cultural practice on the population leaves of* Brachycolus asparagi *in spring (after Halfhill et al. 1984)*

	Aphids/spear		Infested spears	
	No tillage	Spring tillage	No tillage	Spring tillage
Untreated	4.6	0.8	46	26
Autumn burn	1.6	0.3	32	11
Autumn remove	1.5	0.2	40	7
Autumn mow	1.1	0.1	23	7

The greenbug, Schizaphis graminum, *in the United States*

The greenbug, *Schizaphis graminum* (Rondani) (Homoptera: Aphididae), is a serious pest of cereals in the United States. However, they are susceptible to low overnight temperatures as they overwinter as active nymphs feeding on graminaceous plants. If cattle are allowed to graze in fields infested with *S. graminum*, even if only for a few days, a 51 per cent decrease in aphid numbers can result. Continuous grazing can result in a 98 per cent reduction in aphid numbers and subsequent prevention of greenbug damage to the crop during the growing season (Arnold 1981). It has been suggested that not only could this be used as a control method for this insect but that a knowledge of the weather and grazing conditions in a particular area could be useful in predicting greenbug numbers.

Conclusion

Thus, cultural practice can be a very good method of controlling insect pests through the overwintering stage. However, not all insects are amenable to this means of control; much depends on climate. For example, the corn rootworms, *Diabrotica longicornis* (Say) and *D. virgifera* Le Conte (Coleoptera: Chrysomelidae), both overwinter as eggs in the soil in South Dakota. Spring and autumn ploughing can cause increased egg mortality by bringing the eggs closer to the surface (Calkins and Kirk 1969), but due to the very variable winters in South Dakota it cannot be recommended as a reliable method of control.

For successful prediction and control of insects based on the overwintering stage, a detailed understanding of the processes involved in overwintering is required. This information is, unfortunately, as this chapter has shown, available for only a very limited number of species. We hope that the preceding examples have highlighted the advantages to be gained by prediction and control at this stage of the insect life cycle and that research will be stimulated in this most productive area.

Bibliography

Alford, D. V. (1969). A study of the hibernation of bumblebees (Hymenoptera: Bombidae) in southern England. *Journal of Animal Ecology*, **38**, 149–70.

Anderson, N. H. (1962). Studies on overwintering of *Anthocoris* (Hem., Anthocoridae). *Entomologist's Monthly Magazine*, **98**, 1–3.

Anderson, J. F. and Kaya, H. K. (1974). Diapausal induction by photoperiod and temperature in the elm spanworm egg parasitoid, *Ooencyrtus* sp. *Annals of the Entomological Society of America*, **67**, 845–9.

Arnold, D. C. (1981). Effects of cold temperature and grazing on greenbug populations in wheat in Noble County, Oklahoma, 1975–76. *Journal of the Kansas Entomological Society*, **54**, 571–7.

Arthur, A. P. and Mason, P. G. (1985). The life history and immature stages of *Banchus flavescens* (Hymenoptera: Ichneumonoidae), a parasitoid of the Bertha Armyworm, *Mamestra configurata* (Lepidoptera: Noctuidae) in Western Canada. *Canadian Entomologist*, **117**, 1249–55.

Asahina, E. (1966). Freezing and frost resistance in insects. In *Cryobiology*, (ed. H. T. Meryman), pp. 451–86. Academic Press, London.

Asahina, E. (1969). Frost resistance in insects. *Advances in Insect Physiology*, **6**, 1–49.

Ashwood-Smith, M. J. (1970). In *Current trends in cryobiology* (ed. A. U. Smith), pp. 5–42. Plenum Press, New York.

Askew, R. R. (1971). *Parasitic insects*. Elsevier, New York.

Austarä, Ø. (1971). Cold-hardiness in eggs of *Neodiprion sertifer* (Geoffroy) (Hym., Diprionidae) under natural conditions. *Norsk Entomologist Tidsskrift*, **18**, 45–8.

Bakke, A. (1963). Studies on the spruce-cone insects *Laspeyresia strobilella* (L) (Lepidoptera: Tortricidae), *Kaltenbachiola strobi* (Winn.) (Diptera: Itonididae) and their parasites (Hymenoptera) in Norway. *Report of the Norwegian Forest Research Institute*, **67**, 1–151.

Bakke, A. (1969). Extremely low supercooling point in eggs of *Zeiraphera diniana* (Guenee) (Lepidoptera: Tortricidae). *Norwegian Journal of Entomology*, **16**, 81–3.

Bakke, A. (1971). Distribution of prolonged diapausing larvae in populations of *Laspeyresia strobilella* L. (Lep., Tortricidae) from spruce cones. *Norsk Entomologist Tidsskrift*, **18**, 89–93.

Bale, J. S. (1979). The occurrence of an adult reproductive diapause in the univoltine life cycle of the beech leaf mining weevil, *Rhynchaenus fagi* L. *International Journal of Invertebrate Reproduction*, **1**, 57–66.

Bale, J. S. (1980). Seasonal variation in cold hardiness of the adult beech leaf mining weevil *Rhynchaenus fagi* L. in Great Britain. *Cryo-Letters*, **1**, 372–83.

Bale, J. S. (1987). Review – Insect cold hardiness: freezing and supercooling – an ecophysiological perspective. *Journal of Insect Physiology*, **33**, 899–908.

Bale, J. S. (1989). Cold hardiness and overwintering survival of insects. *Agricultural Zoology Reviews*, **3**, 157–92.

Bale, J. S. and Pullin, A. S. (1991). Opportunities and risks in the overwintering strategy of a wall-dwelling species of *Hypogastrura* (collembola). *Cryo-Letters*, **12**, 155–162.

Bale, J. S. and Smith, G. C. (1981). The nature of cold acclimatisation in the beech leaf mining weevil *Rhynchaenus fagi* L. *Cryo-Letters*, **2**, 325–36.

Bale, J. S., O'Doherty, R., Atkinson, H. J. and Stevenson, R. A. (1984). An automatic thermoelectric cooling method and computer-based recording system for supercooling point studies on small invertebrates. *Cryobiology*, **21**, 340–7.

Bale, J. S., Harrington, R. and Clough, M. S. (1987). Effect of low temperature on the survival of the peach-potato aphid *Myzus persicae*. *Proceedings of the Third European Congress of Entomology*, 243–246. Amsterdam 1986.

Bale, J. S., Harrington, R. and Clough, M. S. (1988). Low temperature mortality of the peach-potato aphid *Myzus persicae*. *Ecological Entomology*, **13**, 121–9.

Bale, J. S., Hansen, T. N. and Baust, J. G. (1989a) Nucleators and sites of nucleation in the freeze tolerant larvae of the gall fly *Eurosta solidaginis* (Fitch). *Journal of Insect Physiology*, **35**, 291–8.

Bale, J. S., Hansen, T. N., Nishino, M. and Baust, J. G. (1989b). Effect of cooling rate on the survival of larvae, pupariation and adult emergence of the gall fly *Eurosta solidaginis*. *Cryobiology*, **26**, 285–9.

Barbour, D. A. (1985). Patterns of population fluctuation in the pine looper moth *Bupalus piniaria* L. in Britain. In *Site characteristics and population dynamics of Lepidopteran and Hymenopteran pests*, Forestry Commission Research and Development Paper **135**, (ed. D. Bevan and J. T. Stoakley), pp. 8–20.

Barbour, D. A. (1986). Expansion of range of the speckled wood butterfly, *Pararge aegeria* L. in north-east Scotland. *Entomologist's Record and Journal of Variation*, **98**, 98–105.

Baronio, P. and Sehnal, F. (1980). Dependence of the parasitoid *Conia cinerascens* on the hormones of its Lepidopterous hosts. *Journal of Insect Physiology*, **26**, 619–26.

Barry, R. G. and Chorley, R. J. (1976). *Atmosphere, weather and climate*, (3rd edn). Methuen, London.

Barson, G. (1974). Some effects of freezing temperatures on overwintering larvae of the large elm bark beetle (*Scolytus scolytus*). *Annals of Applied Biology*, **78**, 219–24.

Basedow, T. (1977). Der Einfluss von Temperatur und Niederschlägen auf Diapause und Phänologie die Weizengallmücken *Contarinia tritici* (Kirby) und *Sitodiplosis mosellana* (Géhin) (Dipt. Cecidomyidae). *Zoologische Jahrbücher Abteilung für Systematik Ökologie und Geographie der Tiere*.

Baust, J. G. (1973). Mechanisms of cryoprotection in freezing tolerant animal systems. *Cryobiology*, **10**, 197–205.

Baust, J. G. (1976). Temperature buffering in an arctic microhabitat. *Annals of the Entomological Society of America,* **69**, 117–20.

Baust, J. G. (1980). Low temperature tolerance in an Antarctic insect: a relict adaptation? *Cryo-Letters,* **1**, 360–71.

Baust, J. G. (1981). Biochemical correlates to cold hardening in insects. *Cryobiology,* **18**, 186–98.

Baust, J. G. (1982). Environmental triggers to cold hardening. *Comparative Biochemistry and Physiology,* **73A**, 563–70.

Baust, J. G. and Edwards, J. S. (1979). Mechanisms of freezing tolerance in an Antarctic midge, *Belgica antarctica. Physiological Entomology,* **4**, 1–5.

Baust, J. G. and Lee, R. E. (1981). Divergent mechanisms of frost hardiness in two populations of gall fly, *Eurosta solidaginis. Journal of Insect Physiology,* **27**, 485–90.

Baust, J. G. and Miller, L. K. (1970). Variations in glycerol content and its influence on cold hardiness in the Alaskan Carabid beetle *Pterostichus brevicornis. Journal of Insect Physiology,* **16**, 979–90.

Baust, J. G. and Miller, L. K. (1972). Influence of low temperature acclimation on cold hardiness in the beetle, *Pterostichus brevicornis. Journal of Insect Physiology,* **18**, 1935–47.

Baust, J. G. and Morrisey, R. E. (1977). Strategies of low temperature adaptation. *Proceedings of the XVth International Congress of Entomology,* Montreal, 173–184.

Baust, J. G. and Rojas, R. R. (1985). Review – Insect cold hardiness: facts and fancy. *Journal of Insect Physiology,* **31**, 755–9.

Baust, J. G. and Zachariassen (1983). Seasonally active cell matrix associated ice nucleators in an insect. *Cryo-Letters,* **4**, 65–71.

Baxendale, F. P. and Teetes, G. L. (1983). Factors influencing adult emergence from diapausing sorghum midge, *Contarinia sorghicola* (Diptera: Cecidomyiidae). *Environmental Entomology,* **12**, 1064–7.

Bean, D. W. and Beck, S. D. (1983). Haemolymph ecdysteroid titres in diapause and non-diapause larvae of the European corn borer, *Ostrinia nubilalis. Journal of Insect Physiology,* **81**, 687–93.

Beck, S. D. (1962). Temperature effects on insects: relation to periodism. *Proceedings of the North Central Branch of the Entomological Society of America,* **17**, 18–19.

Beck, S. D. (1967). Water intake and the termination of diapause in the European corn borer, *Ostrinia nubilalis. Journal of Insect Physiology,* **13**, 739–50.

Beck, S. D. (1980). *Insect photoperiodism,* (2nd edn). Academic Press, New York.

Beck, S. D. (1982). Thermoperiodic induction of larval diapause in the European corn borer, *Ostrinia nubilalis. Journal of Insect Physiology,* **28**, 273–7.

Beck, S. D. (1983a). Insect thermoperiodism. *Annual Review of Entomology,* **28**, 91–108.

Beck, S. D. (1983b). Thermal and thermoperiodic effects on larval development and diapause in the European corn borer *Ostrinia nubilalis. Journal of Insect Physiology,* **29**, 107–12.

Beck, S. D. (1988). Thermoperiod and larval development of *Agrotis ipsilon* (Lepidoptera: Noctuidae). *Annals of the Entomology Society of America,* **81**, 831–5.

Beckwith, R. C. (1976). Influence of host foliage on the Douglas-fir tussock moth. *Environmental Entomology,* **5**, 73–7.

Beckwith, R. C., Mason, R. R. and Paul, H. G. (1978). Regression for estimating numbers of Douglas-fir tussock moth eggs relative to egg mass weight. *Canadian Entomologist*, **110**, 131–4.

Bell, A. C. (1983). The life-history of the leaf-curling plum aphid *Brachycaudus helichrysi* in northern Ireland and its ability to transmit potato virus Yc(AB). *Annals of Applied Biology*, **102**, 1–6.

Bell, C. H. (1976). Factors governing the induction of diapause in *Ephestia eluntella* and *Plodia interpunctella* (Lepidoptera). *Physiological Entomology*, **1**, 83–91.

Benham, G. S. and Farrar, R. J. (1976). Notes on the biology of *Prionus laticollis* (Coleoptera: Cerambycidae). *Canadian Entomologist*, **108**, 569–76.

Bennett, L. E. and Lee, R. E. (1989). Simulated winter to summer transition in diapausing adults of the lady beetle (*Hippodamia convergens*): supercooling point is not indicative of cold-hardiness. *Physiological Entomology*, **14**, 361–7.

Bevan, D. and Brown, R. M. (1978). *Pine looper moth*. Forestry Commission Record 119, HMSO, London.

Biernaux, J. (1968). Observations sur l'hibernation de *Psila rosae* F. *Bulletin Recherches Agronomiques de Gembloux*, **3**, 241–8.

Birch, L. C. (1942). The influence of temperatures above the developmental zero on the development of the eggs of *Austroicetes cruciata* Sauss. (Orthoptera). *Australian Journal of Experimental Biology and Science*, **20**, 17–25.

Blais, J. R. (1960). Spruce budworm parasite investigations in the low St Lawrence and Caspe regions of Quebec. *Canadian Entomologist*, **92**, 384–96.

Block, W. (1982a). Cold hardiness in invertebrate poikilotherms. *Comparative Biochemistry and Physiology*, **73A**, 581–93.

Block, W. (1982b). Supercooling points of insects and mites on the Antarctic Peninsula. *Ecological Entomology*, **7**, 1–8.

Block, W. (1982c). The Signy Island Terrestrial Reference Sites: XIV. Population studies on the *Collembola*. *British Antarctic Survey Bulletin*, **55**, 33–49.

Block, W. (1987). Temperature and drought effects on Antarctic land arthropods. *Proceedings of the Third European Congress of Entomology*, 247–50. Amsterdam, 1986.

Block, W. and Young, S. R. (1979). Measurement of supercooling in small arthropods and water droplets. *Cryo-Letters*, **1**, 85–91.

Block, W. and Zettel, J. (1980). Cold hardiness of some alpine *Collembola*. *Ecological Entomology*, **5**, 1–9.

Block, W., Turnock, W. J. and Jones, T. H. (1987). Cold resistance and overwintering survival of the cabbage root fly, *Delia radicum* (Anthomyiidae), and its parasitoid, *Trybliographa rapae* (Cynipidae), in England. *Oecologia*, **71**, 332–8.

Bogenschutz, H. (1976). Untersuchungen über den Einfluss der Temperatur auf die Entwicklung von *Rhyacionia buoliana* Den. u. Schiff. (Lep., Tortricidae). *Zeitschrift für Pflanzenkultur und Pflanzenschutz*, **83**, 22–39.

Bohm, M. K. (1972). Effects of environment and juvenile hormone on ovaries of the wasp, *Polistes metricus*. *Journal of Insect Physiology*, **18**, 1875–83.

Bonnemaison, L. (1951). Contribution a l'étude des facteurs provoquant l'apparition des formes ailées et sexuées chez les aphidinae. *Annales de Epiphyties* (C), **2**, 1–380.

Botella, L. M. and Mensua, J. L. (1987). Larval diapause induced by crowding in *Chymomyza costata* (Diptera: Drosophilidae). *Annales Entomologici Fennici*, **53**, 41–7.

Bournoville, R. (1973). Observations écologiques sur l'hivernation du puceron du pois *Acyrthosiphon pisum* (Harris) et de ses parasites dans la région de Versailles. *Annales Zoologie-Ecologie d'Animaux*, **5**, 13–28.

Bradshaw, W. E. (1970). Interaction of food and photoperiod in the termination of larval diapause in *Chaoborus americanus*. *Biological Bulletin*, **139**, 476–84.

Bradshaw, W. E. (1976). Geography of photoperiodic response in a diapausing mosquito. *Nature*, **262**, 384–5.

Bradshaw, W. E. and Holzappfel, C. M. (1983). Life cycle strategies in *Wyeomyia smithii*: seasonal and geographical adaptations. In *Diapause and life cycle strategies in insects* (ed. V. K. Brown and I. Hodek), pp. 167–85. Dr W. Junk Publishers, The Hague.

Bradshaw, W. E. and Lounibos, L. P. (1972). Photoperiodic control of development in the pitcher-plant mosquito, *Wyeomyia smithii*. *Canadian Journal of Zoology*, **50**, 713–19.

Brakefield, P. M. (1985). Differential winter mortality and seasonal selection in the polymorphic ladybird *Adalia bipunctata* (L.) in the Netherlands. *Biological Journal of the Linnean Society*, **24**, 189–206.

Brian, M. V. (1979). Caste differentiation and division of labor. In *Social insects*, vol. 1, (ed. H. R. Hermann), pp. 121–2. Academic Press, New York.

Brian, M. V. and Kelly, A. F. (1967). Studies of caste differentiation in *Myrmica rubra* L.9. Maternal environment and the caste bias of larvae. *Insects sociaux*, **14**, 13—24.

Brodeur, J. and McNeil, J. N. (1989). Biotic and abiotic factors involved in diapause induction of the parasitoid, *Aphidius nigripes* (Hymenoptera: Aphidiidae). *Journal of Insect Physiology*, **35**, 969–74.

Bronson, T. E. (1935). Observations on winter survival of pea aphid eggs. *Journal of Economic Entomology*, **28**, 1030–6.

Brunel, E. and Missonnier, J. (1968). Etude du développement nymphal de *Psila rosae* Fab. (Dipteres Psilides) en conditions naturelles et experimentales: quiescenel et diapause. *C.r. Seane. Soc. Biol.*, **162**, 2223–8.

Burn, A. J. (1984). Life cycle strategies in two Antarctic Collembola. *Oecologia*, **64**, 223–9.

Burn, A. J. and Coaker, T. H. (1981). Diapause and overwintering of the carrot fly, *Psila rosae* (F.) (Diptera: Psilidae). *Bulletin of Entomological Research*, **71**, 583–90.

Butterfield, J. (1976). Effect of photoperiod on a winter and on a summer diapause in two species of cranefly (Tipulidae). Journal of Insect Physiology, **22**, 1443–6.

Cadman, C. H. and Chambers, J. (1960). Factors affecting the spread of aphid-borne viruses in potato in eastern Scotland. III Effects of planting date, roguing and age of crop on the spread of potato leaf-roll and Y viruses. *Annals of Applied Biology*, **48**, 729–38.

Caldwell, E. T. N. and Wright, R. E. (1978). Induction and termination of diapause in the face fly, *Musca autumnalis* (Diptera: Muscidae), in the laboratory. *Canadian Entomologist*, **110**, 617–22.

Calkins, C. O. and Kirk, V. M. (1969). Effect of winter precipitation and temperature on overwintering eggs of northern and western corn rootworms. *Journal of Economic Entomology*, **62**, 541–3.

Calvert, W. H., Zuchowski, W. and Brower, L. P. (1983). The effect of rain, snow and freezing temperatures on overwintering monarch butterflies in Mexico. *Biotropica*, **15**, 42–7.

Cammell, M. E., Way, M. J. and Heathcote, G. D. (1978). Distribution of eggs of the black bean aphid, *Aphis fabae* Scop., on the spindle bush, *Euonymus europaeus L.*, with reference to forecasting infestations of the aphid on field beans. *Plant Pathology*, **27**, 68–76.

Cannon, R. J. C. (1986a). Effects of ingestion of liquids on the cold tolerance of an Antarctic mite. *Journal of Insect Physiology*, **32**, 955–61.

Cannon, R. J. C. (1986b). Effects of contrasting relative humidities on the cold tolerance of an Antarctic mite. *Journal of Insect Physiology*, **32**, 523–34.

Cannon, R. J. C. (1986c). Diet and acclimation effects on the cold tolerance and survival of an Antarctic springtail. *British Antarctic Survey Bulletin*, **No. 71**, 19–30.

Cannon, R. J. C. and Block, W. (1988). Cold tolerance of microarthropods. *Biological Review*, **63**, 23–77.

Cannon, R. J. C., Block, W. and Collett, G. D. (1985). Loss of supercooling ability on *Cryptopygus antarcticus* (Collembola: Isotomidae) associated with water uptake. *Cryo-Letters*, **6**, 73–80.

Carante, J. P. and Lemaître, C. (1990). Some responses to simulated winter stresses in adults of the Mediterranean fruit fly (Diptera: Tephritidae). *Annals of the Entomological Society of America*, **83**, 36–42.

Carter, C. I. (1971). Winter temperatures and survival of the green spruce aphid, *Elatobium abietinum* (Walker). *Forest Record*, **84**, 3–10.

Carter, C. I. (1977). *Impact of green spruce aphid on growth*. Forestry Commission Research and Development Paper no. 116, 1–8.

Carter, C. I. and Nichols, J. F. A. (1988). *The green spruce aphid and Sitka spruce provenances in Britain*. Forestry Commission Occasional Paper no. 19, 1–7.

Carter, C. I. and Nichols, J. F. A. (1989). Winter survival of the lupin aphid *Macrosiphum albifrons* Essig. *Journal of Applied Entomology*, **108**, 213–16.

Casagrande, R. A. and Haynes, D. L. (1976). A predictive model for cereal leaf beetle mortality from sub-freezing temperatures. *Environmental Entomology*, **5**, 761–9.

Causse, R. (1976). Étude de la localisation et de la mortalité hivernale des larves de *Laspeyresia pomonella* L. (Lepidoptera, Tortricidae) en vergers modernes de pommiers de la Baste Vallée du Rhône. *Annales Zoologique-Ecologie des Animaux*, **8**, 83–101.

Chambers, R. J. (1982). Maternal experience of crowding and duration of aestivation in the sycamore aphid (*Drepanosiphum platanoidis*). *Oikos*, **39**, 100–2.

Chaplin, S. B. and Wells, P. H. (1982). Energy reserves and metabolic expenditures of monarch butterflies overwintering in southern California. *Ecological Entomology*, **7**, 249–56.

Chapman, R. F. (1971). *The insects, structure and function*. Elsevier, New York.

Chen, C.-P., Denlinger, D. L. and Lee, R. E. (1987). Cold-shock injury and rapid cold hardening in the flesh fly *Sarcophaga crassipalpis*. *Physiological Zoology*, **60**, 297–304.

Chino, H. (1957). Conversion of glycogen to sorbitol and glycerol in the diapause egg of the *Bombyx* silkworm. *Nature*, **180**, 606–7.

Chippendale, G. M. and Reddy, A. S. (1973). Temperature and photoperiodic regulation of diapause of the south-western corn borer, *Diatraea grandiosella*. *Journal of Insect Physiology*, **19**, 1397–408.

Chippendale, G. M., Reddy, A. S. and Catt, C. L. (1976). Photoperiodic and thermoperiodic interactions in the regulation of the larval diapause of *Diatraea grandiosella*. *Journal of Insect Physiology*, **22**, 823–8.

Chroston, J. R. (1982). *Colour variation in the English grain aphid, Sitobion avenae.* PhD Thesis, University of East Anglia, Norwich.

Church, N. S. and Salt, R. W. (1952). Some effects of temperature on development and diapause in eggs of *Melanoplus bivittatus* (Say) (Orthoptera: Acrididae). *Canadian Journal of Zoology*, **30**, 173–84.

Claret, J. (1973). La diapause facultative de *Pimpla instigator* (Hymenoptera, Ichneumonoidea). I. Rôle de la photopériode. *Entomophaga*, **18**, 409–18.

Cloudsley-Thompson, J. L. (1973). Factors influencing the supercooling of tropical Arthropoda, especially locusts. *Journal of Natural History*, **7**, 471–80.

Collier, R. H. and Finch, S. (1983). Completion of diapause in field populations of the cabbage root fly (*Delia radicum*). *Entomologia Experimentalis et Applicata*, **34**, 186–92.

Copp, N. H. (1983). Temperature-dependent behaviours and cluster formation by aggregating ladybird beetles. *Animal Behaviour*, **31**, 424–30.

Coppock, L. J. (1974). Notes on the biology of carrot fly in Eastern England. *Plant Pathology*, **23**, 93–100.

Corbet, P. S. (1956). The influence of temperature on diapause development in the dragonfly *Lestes sponsa* (Hansemann) (Odonata: Lestidae). *Proceedings of the Royal Entomological Society of London*, (A)**31**, 45–8.

Corbet, P. S. and Danks, H. V. (1975). Egg-laying habits of mosquitoes in the high arctic. *Mosquito News*, **35**, 8–14.

Courtin, G. M., Shorthouse, J. D. and West, R. J. (1984). Energy relations of the snow scorpionfly *Boreus brumalis* (Mecoptera) on the surface of snow. *Oikos*, **43**, 241–5.

Crowe, J. and Clegg, J. S. (1973). *Anhydrobiosis.* Dooden, Hutchinson and Ross, Stroudsbery, Pennsylvania.

Danilevskii, A. S. (1961). *Photoperiodism and seasonal development of insects.* Oliver and Boyd, London (1965 English translation).

Danilevskii, A. S. (1965). *Photoperiodism and seasonal development of insects.* (English edn), Oliver and Boyd, Edinburgh and London.

Danilevskii, A. S., Goryshin, N. I. and Tyshenka, V. P. (1970). Biological rhythms in terrestrial arthropods. *Annual Review of Entomology*, **15**, 201–44.

Danks, H. V. (1978). Modes of seasonal adaptation in insects. 1. Winter survival. *Canadian Entomologist*, **110**, 1167–205.

Danks, H. V. (1987). *Insect dormancy: an ecological perspective.* Biological Survey of Canada, Ottawa.

Danks, H. V. and Byers, J. R. (1972). Insects and arachnids of Bathurst Island, Canadian arctic archipelago. *Canadian Entomologist*, **104**, 81–8.

Dasch, C. E. (1971). Hibernating Ichneumonidae of Ohio (Hymenoptera). *The Ohio Journal of Science*, **71**, 270–83.

Dean, G. J. (1974). The overwintering and abundance of cereal aphids. *Annals of Applied Biology*, **76**, 1–7.

Dean, R. L. and Hartley, J. C. (1977). Egg diapause in *Ephippiger cruciger* (Orthoptera: Tettigoniidae) II. The intensity and elimination of the final egg diapause. *Journal of Experimental Biology*, **66**, 185–95.

Denlinger, D. L. (1972). Seasonal phenology of diapause in the flesh fly *Sarcophaga bullata. Annals of the Entomological Society America*, **65**, 410–14.

Denlinger, D. L. and Bradfield, J. Y. (1981). Duration of pupal diapause in the tobacco hornworm is determined by the number of short days received by the larva. *Journal of Experimental Biology*, **91**, 331–7.

Deseo, K. V. (1973). Reproductive activity of codling moth (*Laspeyresia pomonella* L. Lepidopt.; Tortr.) exposed to short photophase during preimaginal state. *Acta Phytopathologica Hungarica*, **8**, 193–206.

Deseo, K. V. and Briolini, G. (1986). Observations on the termination of the facultative diapause in the codling moth (*Cydia pomonella* L., Lepidoptera: Tortricidae). *Bollettino dell'Istituto di Entomologia 'Guido Grandi' della Università degli studi di Bologna*, **40**, 99–110.

Deseo, K. V. and Saringer, G. (1975). Photoperiodic effect on fecundity of *Laspeyresia pomonella, Grapholitha funebrana* and *G. molesta*: the sensitivity period. *Entomologia experimentalis et Applicata*, **18**, 187–93.

Deura, K. and Hartley, J. C. (1982). Initial diapause and embryonic development in the speckled bush-cricket, *Leptophytes punctatissima*. *Physiological Entomology*, **7**, 253–62.

DeVries, A. L. (1980). Biological antifreezes and survival in freezing environments. In *Animals and environmental fitness*, (ed R. Giles), pp. 583–607). Pergamon, New York.

DeVries, A. L. (1982). Biological antifreeze agents in cold water fishes. *Comparative Biochemistry and Physiology*, **73A**, 627–40.

Dewar, A. M. and Carter, N. (1984). Decision trees to assess the risk of cereal aphid outbreaks. *Bulletin of Entomological Research*, **74**, 387–98.

De Wilde, J., Duintjer, C. S. and Mook, L. (1959). Physiology of diapause in the adult Colorado beetle (*Leptinotarsa decemlineata* Eng). 1. The photoperiod as controlling factor. *Journal of Insect Physiology*, **3**, 75–85.

Dixon, A. F. G. (1966). The effect of population density and nutritive status of the host on the summer reproductive activity of the sycamore aphid, *Drepanosiphum platanoides* (Schr.). *Journal of Animal Ecology*, **35**, 105–12.

Dixon, A. F. G. (1975). Seasonal changes in fat content, form, state of gonads and length of adult life in the sycamore aphid, *Drepanosiphum platanoides* (Schr). *Transactions of the Royal Entomological Society of London*, **127**, 87–99.

Dixon, A. F. G. (1976). Reproductive strategies of the alate morphs of the bird cherry-oat aphid *Rhopalosiphum padi*. (L.) *Journal of Animal Ecology*, **45**, 817–30.

Dixon, A. F. G. (1985). *Aphid ecology*. Blackie, London.

Dixon, A. F. G. (1987). Seasonal development in aphids. In *Aphids: their biology, natural enemies and control*, vol A. (ed A. K. Minks and P. Harrewijn), pp. 315–20. Elsevier, Amsterdam.

Dixon, A. F. G. and Glen, D. M. (1971). Morph determination in the bird cherry-oat aphid, *Rhopalosiphum padi* (L.). *Annals of Applied Biology*, **68**, 11–21.

Downes, J. A. (1962). What is an arctic insect? *Canadian Entomologist*, **94**, 143–63.

Duman, J. G. (1977a). The role of macromolecular antifreeze in the darkling beetle, *Meracantha contracta*. *Journal of Comparative Physiology*, **115**, 279–86.

Duman, J. G. (1977b). Variations in macromolecular antifreeze levels in larvae of the darkling beetle, *Meracantha contracta*. *Journal of Experimental Zoology*, **201**, 85–92.

Duman, J. G. (1977c). Environmental effects on antifreeze levels in larvae of the darkling beetle, *Meracantha contracta*. *Journal of Experimental Zoology*, **201**, 333–7.

Duman, J. G. (1979). Thermal hysteresis factors in overwintering insects. *Journal of Insect Physiology*, **25**, 805–10.

Duman, J. G. (1980). Factors involved in the overwintering survival of the freeze tolerant beetle *Dendroides canadensis*. *Journal of Comparative Physiology*, **136**, 53–9.

Duman, J. G. (1982). Insect antifreezes and ice nucleating agents. *Cryobiology*, **19**, 613–27.

Duman, J. G. and Horwath, K. L. (1983). The role of haemolymph proteins in the cold tolerance of insects. *Annual Review of Physiology*, **45**, 261–70.

Duman, J. G. and Patterson, J. L. (1978). The role of ice nucleators in the frost tolerance of overwintering queens of the bald faced hornet. *Comparative Biochemistry and Physiology*, **59A**, 69–72.

Duman, J. G., Morris, J. P. and Castellino, F. J. (1984). Purification and composition of an ice nucleating protein from queens of the hornet, *Vespula maculata. Journal of Comparative Physiology*, **B.154**, 79–83.

Duman, J. G., Horwath, K. L., Tomchaney, A. and Patterson, J. L. (1982). Antifreeze agents of terrestrial arthropods. *Comparative Biochemistry and Physiology*, **73A**, 545–55.

Dunn, J. A. (1959). The survival in soil of apterae of lettuce root aphid. *Annals of Applied Biology*, **47**, 766–71.

Dunn, J. A. and Wright, D. W. (1955). Overwintering egg populations of the pea-aphid in East Anglia. *Bulletin of Entomological Research*, **46**, 389–92.

Edmunds, G. F. (1973). Ecology of black pineleaf scale (Homoptera: Diaspididae). *Environmental Entomology*, **2**, 765–77.

Eguagie, W. E. (1974). Cold hardiness of *Tingis ampliata* (Heteroptera: Tingidae). *Entomologia Experimentalis et Applicata*, **17**, 204–14.

Eidt, D. C. and Little, C. H. A. (1968). Insect control by artificially prolonging plant dormancy – a new approach. *Canadian Entomologist*, **100**, 1278–9.

Elbert, A. (1979). Ökologische Untersuchungen zur Steuerung der Lanraldiapause von *Trogderma variabile* Ballion (Col. Dermestidae). *Zeitschrift für Angewandte Entomologie*, **88**, 268–82.

Embree, D. G. (1967). Effects of the winter moth on growth and mortality of red oak in Nova Scotia. *Forest Science*, **13**, 295–9.

Enomoto, O. (1981). Larval diapause in *Chymomyza costata* (Diptera: Drosophilidae). I. Effects of temperature and photoperiod on the development. *Low Temperature Science Series*, **B39**, 21–9.

Eskafi, F. M. and Legner, E. F. (1974). Fecundity, development and diapause in *Hexacala* sp. near *websteri* a parasite of *Hippelates* eye gnats. *Annals of the Entomological Society of America*, **67**, 769–71.

Fields, P. G. and McNeil, J. N. (1988). Characteristics of the larval diapause in *Ctenucha virginica* (Lepidoptera: Arctiidae). *Journal of Insect Physiology*, **34**, 111–15.

Fisken, A. G. (1959). Factors affecting the spread of aphid-borne viruses in potato in eastern Scotland. I. Overwintering of potato aphids, particularly *Myzus persicae* (Sulzer). *Annals of Applied Biology*, **47**, 274–86.

Flint, M. L. (1980). Climatic ecotypes in *Trioxys complanatus*, a parasite of the spotted alfalfa aphid. *Environmental Entomology*, **9**, 501–7.

Flohn, H. (1969). *Climate and weather*. Weidenfeld and Nicolson, London.

Ford, E. B. (1945). *Butterflies*. Collins, London.

Forrest, J. M. S. (1970). The effects of maternal and larval experience on morph determinations in *Dysaphis devecta. Journal of Insect Physiology*, **16**, 2281–92.

Frankos, V. H. and Platt, A. P. (1976). Glycerol accumulation and water content in larvae of *Limenitis archippus*: importance in winter survival. *Journal of Insect Physiology*, **22**, 632–8.

Frazer, B. D. and Raworth, D. A. (1984). Predicting the time of hatch of the strawberry aphid, *Chaetosiphon fragaefolii* (Homoptera: Aphididae). *Canadian Entomologist*, **116**, 1131–5.

Fuchs, T. W. and England, A. (1989). Winter habitat sampling for overwintering boll weevils (Coleoptera: Curculionidae) as a component of an integrated pest management program. *Southwestern Entomologist*, 14, 265–70.

Fuller, W. A., Stebbins, L. L. and Dyke, G. R. (1969). Overwintering of small mammals near Great Slave Lake, Northern Canada. *Arctic*, 22, 34–55.

Furunishi, S. and Masaki, S. (1981). Photoperiodic response of the univoltine ant-lion *Myrmeleon frinicorius* (Neuroptera, Myrmeleontidae). *Kontyû*, 49, 653–67.

Gaines, R. C. (1953). Relation between winter temperatures, boll weevil survival, summer rainfall and cotton yields. *Journal of Economic Entomology*, 46, 685–8.

Gangavalli, R. R. and AliNiazee, M. T. (1985). Diapause induction in the oblique banded leafroller *Chiristoneura rosaceana* (Lepidoptera: Tortricidae): role of photoperiod and temperature. *Journal of Insect Physiology*, 31, 831–5.

Gange, A. (1989). Overwintering in the birch aphid, *Euceraphis punctipennis*. *British Journal of Entomology and Natural History*, 2, 181–3.

Gange, A. C. and Llewellyn, M. (1988). Egg distribution and mortality in the alder aphid, *Pterocallis alni*. *Entomologia Experimentalis et Applicata*, 48, 9–14.

Gara, R. I. and Wood, J. Q. (1989). Termination of reproductive diapause in the Sitka spruce weevil, *Pissodes strobi* (Peck) (Col., Curculionidae) in western Washington. *Journal of Applied Entomology*, 108, 156–63.

Garcia-Salazar, C., Podoler, H. and Wholan, M. E. (1988). Effects of temperature on diapause induction in the codling moth, *Cydia pomonella* (L.) (Lepidoptera: Olethreutidae). *Environmental Entomology*, 17, 626–8.

Gash, A. F. and Bale, J. S. (1985). Host plant influences on supercooling ability of the black-bean aphid *Aphis fabae*. *Cryo-Letters*, 6, 297–304.

Gehrken, U. (1984). Winter survival of an adult bark beetle *Ips acuminatus* Gyll. *Journal of Insect Physiology*, 30, 421–9.

Gelman, D. B. and Woods, C. W. (1983). Haemolymph ecolysteroid titers of diapause and non-diapause bound fifth instars and pupae of the European corn borer *Ostrinia nubilalis* (Hübner). *Comparative Biochemistry and Physiology*, 76A, 367–75.

Gerber, G. H. (1984). Influence of date of oviposition on egg hatching and embryo survival in the red turnip beetle, *Entomoscelis americana* (Coleoptera: Chrysomelidae). *Canadian Entomologist*, 116, 645–52.

Gerber, G. H. and Lamb, R. J. (1982). Phenology of egg hatching for the red turnip beetle *Entomoscelis americana* (Coleoptera: Chrysomelidae). *Environmental Entomology*, 11, 1258–63.

Gerhold, H. D. (1966). In quest of insect-resistant forest trees. In *Breeding for resistant trees*, (eds H. D. Gerhold, E. J. Schreiner, R. E. McDermott and J. E. Winleski), pp. 305–18. Pergamon Press, Oxford.

Geri, C. and Goussard, F. (1988). Incidence de la photophase et de la température sur la diapause de *Diprion pini* L. (Hym., Diprionidae). *Journal of Applied Entomology*, 106, 150–72.

Geri, C. and Goussard, F. (1989a). Incidence de la qualité de la lumière sur la développement et la diapause de *Diprion pini* L. (Hym., Diprionidae). *Journal of Applied Entomology*, 108, 89–101.

Geri, C. and Goussard, F. (1989b). Incidence de l'importance numérique des colonies larvaires sur l'induction de la diapause de *Diprion pini* L. (Hym., Diprionidae). *Journal of Applied Entomology*, 108, 131–7.

Geri, C., Goussard, F., Allais, J. P. and Buratti, L. (1988). Incidence de l'alimentation sur le développement et la diapause de *Diprion pini* (Hym. Diprionidae). *Journal of Applied Entomology*, **106**, 451–64.

Greenbank, D. O. (1970). Climate and the ecology of the balsam woolly aphid. *Canadian Entomologist*, **102**, 546–78.

Grenier, S. and Delobel, B. (1984a). Déterminisme de l'arrêt de développement larvaire du parasitoid *Pseudopterichaeta insidiosa* (Diptera: Tachinidae) dans *Galleria mellanella* (Lepidoptera: Pyralidae). *Acta Oecologica, Oecologia Applicata*, **5**, 212–19.

Grenier, S. and Delobel, B. (1984b). Croissance pondérole et arrêt du développement larvaire du parasitoid *Pseudopterichaeta insidiosa* (Diptera: Tachinidae) dans *Galleria mellonella* (Lepidoptera: Pyralidae). *Acta Oecologica Oecologia Applicata*, **5**, 145–52.

Griffiths, E. and Wratten, S. D. (1979). Intra- and interspecific differences in cereal aphid low-temperature tolerance. *Entomologia Experimentalis et Applicata*, **26**, 161–7.

Griffiths, K. J. (1969). Development and diapause in *Pleolophus basizonus* (Hymenoptera: Ichneumonidae). *Canadian Entomologist*, **101**, 907–14.

Grobler, J. H. (1962). The life history and ecology of the woolly pine needle aphid, *Schizolachnus piniradiatae* (Davidson). *Canadian Entomologist*, **94**, 35–45.

Guzman, D. R. and Petersen, J. J. (1986). Overwintering of filth fly parasites (Hymenoptera: Pteromalidae) in open silage in eastern Nebraska. *Environmental Entomology*, **15**, 1296–300.

Hagen, K. S. (1962). Biology and ecology of predacious Coccinellidae. *Annual Review of Entomology*, **7**, 289–326.

Hagstrum, D. W. and Silhacek, D. L. (1980). Diapause induction in *Ephestia cautella*: an interaction between genotype and crowding. *Entomologia Experimentalis et Applicata*, **28**, 29–37.

Halfhill, J. A., Gefre, J. A. and Tamaki, G. (1984). Cultural practice inhibiting overwintering survival of *Brachycaudus asparagi* Mordvilko (Homoptera: Aphididae). *Journal of Economic Entomology*, **77**, 954–6.

Hammer, O. (1942). Biological and ecological investigations of flies associated with pasturing cattle and their excrement. *Videnskabelige Meddelelser fra Dansk naturhistorisk Foreningi Kjøbenhavn*, **105**, 141–393.

Hand, S. C. (1983). The effect of temperature and humidity on the duration of development and hatching success of eggs of the aphid *Sitobion avenae*. *Entomologia experimentalis et applicata*, **33**, 220–222.

Hand, S. C. and Hand, L. (1986). Monitoring of the winter populations of cereal aphids near Wageningen, the Netherlands in 1982/1983. *Netherlands Journal of Plant Pathology*, **92**, 137–46.

Hanover, J. W. (1975). Physiology of tree resistance to insects. *Annual Review of Entomology*, **20**, 75–95.

Hansen, T. (1973). Variations in glycerol content in relation to cold-hardiness on the larvae of *Petrova resinella* L. (Lepidoptera: Tortricidae). *Eesti NSV Teaduste Akadeemia Tometised Bioloogia*, **22**, 105–11.

Hansen, T. (1975). On the resistance of *Mamestra persicariae* L. pupae to cold duration. *Eesti NSV Teaduste Akadeemia Tometised Bioloogia*, **24**, 289–94.

Hanski, I. (1988). Four kinds of extra long diapause in insects: a review of theory and observations. *Annales Zoologici Fennici*, **25**, 37–53.

Hanski, I. (1989). Fungivory: fungi, insects and ecology. In *Insect–fungus interactions*, 14th Symposium of the Royal Entomological Society of

London, (ed. N. Wilding, N. M. Collins, P. M. Hammond and J. F. Webber), pp. 25–68. Academic Press, London.

Hanski, I. and Ståhls, G. (1990). Prolonged diapause in fungivorous *Pegomya* flies. *Ecological Entomology*, **15**, 241–4.

Harcourt, D. G. and Cass, L. M. (1966). Photoperiodism and fecundity in *Plutella maculipennis* (Curt.). *Nature*, **210**, 217–18.

Hare, J. D. (1983). Seasonal variation in plant–insect associations: utilization of *Solanum dulcamara* by *Leptinotarsa decemlineata*. *Ecology*, **64**, 345–61.

Harman, D. M. and Kulman, H. M. (1966). A technique for sexing live white-pine weevils, *Pissodes strobi*. *Annals of the Entomological Society of America*, **59**, 315–17.

Harper, A. M. and Blakeley, P. E. (1968). Survival of *Rhopalosiphum padi* in stored grain during cold weather. *Journal of Economic Entomology*, **61**, 1455.

Harper, A. M. and Lilly, C. E. (1982). Aggregations and winter survival in southern Alberta of *Hippodamia quinquesignata* (Coleoptera: Coccinellidae), a predator of the pear aphid (Homoptera: Aphididae). *Canadian Entomologist*, **114**, 303–9.

Harrington, R. and Cheng, Xia-Nian (1984). Winter mortality, development and reproduction in a field population of *Myzus persicae* (Sulz.) in England. *Bulletin of Entomological Research*, **74**, 633–40.

Havelka, J. (1980). Photoperiodism of the carnivorous midge *Aphidoletes aphidimyza* (Diptera, Cecidomyiidae). *Entomological Review*, **59**, 1–8.

Hayakeva, Y. and Chino, H. (1981). Temperature-dependent interconversion between glycerol and trehalase in diapausing pupae of *Philosamia cynthia, ricini* and *preyeri*. *Insect Biochemistry*, **11**, 43–7.

Hayakeva, Y. and Chino, H. (1982a). Temperature-dependent activation or inactivation of glycogen phosphorylase and synthase of fat body of the silkworm *Philosamia cynthia*: the possible mechanism of the temperature-dependent interconversion between glycogen and trehalase. *Insect Biochemistry*, **12**, 361–6.

Hayakeva, Y. and Chino, H. (1982b). Phosphofructokinase as a possible key enzyme regulating glycerol or trehalase accumulation in diapausing insects. *Insect Biochemistry*, **12**, 639–42.

Hayhoe, H. N. and Mukerji, M. K. (1987). Influence of snow cover on soil temperature in the biologically active zone. Proceedings of the 18th Conference on Agriculture and Forest Meteorology, pp. 45–8. *American Meteorological Society*.

Haynes, D. L., Tummala, R. L. and Ellis, T. L. (1980). Ecosystem management for pest control. *Bioscience*, **30**, 690–6.

Hedlin, A. F., Miller, G. E. and Ruth, D. S. (1982). Induction of prolonged diapause in *Barbara colfasciana* (Lepidoptera: Olethreutidae): correlations with cone crops and weather. *Canadian Entomologist*, **114**, 465–71.

Hegdekar, B. M. (1977). Photoperiodic and temperature regulation of diapause induction in the bertha armyworm *Mamestra configurata*. *Manitoba Entomologist*, **11**, 56–60.

Hegdekar, B. M. (1983). Effect of latitude on the critical photoperiod for diapause induction in the bertha armyworm, *Mamestra configurata* (Lepidoptera: Noctuidae). *Canadian Entomologist*, **115**, 1039–42.

Heinrich, B. (1974). Thermoregulation in endothermic insects. *Science*, **185**, 747–56.

Henson, W. R., Stark, R. W. and Wellington, W. G. (1954). Effects of the weather of the coldest month on winter mortality of the lodgepole needle miner, *Recurvaria* sp. in Banff National Park. *Canadian Entomologist*, **86**, 13–19.

Hille Ris Lambers, D. (1950). De Nederlandse bladluizen van framloos en braam. *Tijdschrift over planteriziekten, Wageningen*, **56**, 253–61.

Hille Ris Lambers, D. (1955). Hemiptera 2. Aphididae. *The zoology of Iceland*, **3, 52a**, 1–29.

Hilton, D. F. J. (1982). The biology of *Endothenia daeckeana* (Lepidoptera: olethreutidae); an inhabitant of the ovaries of the northern pitcher plant *Sarracenia P. purpurea* (Sarraceniaceae). *Canadian Entomologist*, 114, 269–74.

Hochachka, P. W. and Somero, G. N. (1973). *Strategies of Biochemical Adaptation*. W. B. Saunders & Co., Philadelphia.

Hodek, I. (1971a). Termination of adult diapause in *Pyrrhocoris apterus* (Heteroptera: Pyrrhocoridae) in the field. *Entomologia Experimentalis et Applicata*, **14**, 212–22.

Hodek, I. (1971b). Sensitivity to photoperiod in *Aelia acuminata* (L.) after adult diapause. *Oecologia*, **6**, 152–5.

Hodek, I. (1973). *Biology of Coccinellidae*. Dr W. Junk Publishers, The Hague.

Hodek, I. (1975). Diapause development in *Aelia acuminata* females (Heteroptera). *Věstnik Ceskoslovenské Zoologické Spolécnosti*, **39**, 178–94.

Hodek, I. (1976). Two contrasting types of environmental regulation of adult diapause. *Attidella Academia delle Scienze dell Istituto di Bologna*, **13**, 81–8.

Hodek, I. (1979). Intermittent character of adult diapause in *Aelia acuminata* (Heteroptera). *Journal of Insect Physiology*, **25**, 867–71.

Hodek, I. (1983). Role of environmental factors and endogenous mechanisms in the seasonality of reproduction in insects diapausing as adults. In *Diapause and life cycle strategies in insects* (ed. V. K. Brown and I. Hodek), pp. 9–33. Dr W. Junk Publishers, The Hague.

Hodek, I. and Hodková, M. (1988). Multiple role of temperature during insect diapause: a review. *Entomologia Experimentalis et Applicata*, **49**, 153–65.

Hodek, I., Iperti, G. and Rolley, F. (1977). Activation of hibernating *Coccinella septempunctata* (Coleoptera) and *Perilitus coccinellae* (Hymenoptera) and the photoperiodic response after diapause. *Entomologia Experimentalis et Applicata*, **21**, 275–86.

Hodková, M. and Hodek, I. (1987). Photoperiodic summation is temperature-dependent in *Pyrrhocoris apterus* (L.) (Heteroptera). *Experientia*, **43**, 454–6.

Hodková, M., Hodek, I. and Sømme, L. (1989). Cold is not a prerequisite for the completion of photoperiodically induced diapause in *Dolycoris baccarum* from Norway. *Entomologia Experimentalis et Applicata*, **52**, 185–8.

Hodson, A. C. (1937). Some aspects of the role of water in insect hibernation. *Ecological Monographs*, **7**, 271–315.

Holdaway, F. G. and Evans, A. C. (1930). Parasitism a stimulus to pupation: *Alysia manducator* in relation to the host *Lucilia sericata*. *Nature*, **125**, 598–9.

Holmquist, A. M. (1928). Notes on the biology of the muscid fly, *Pyrellia serena* Meigen with special reference to its hibernation. *Annals of the Entomological Society of America*, **21**, 660–7.

Holmquist, A. M. (1931). Studies on arthropod hibernation III. Temperatures in forest hibernacula. *Ecology*, **12**, 387–400.

Hong, J. W. and Platt, A. P. (1975). Critical photoperiod and daylength threshold differences between northern and southern populations of the butterfly *Limenitis archippus*. *Journal of Insect Physiology*, **21**, 1159–65.

Hopkins, A. D. (1938). *Bioclimatics – science of life and climate relations*. US Department of Agriculture, Miscellaneous Publication 280.

Horsfall, W. R., Fowler, H. W., Moretti, L. J. and Larsen, J. R. (1973). *Bionomics and embryology of the inland floodwater mosquito Aedes vexans.* University of Illinois Press, Urbana, Il.

Horwath, K. L. and Duman, J. G. (1982). Involvement of the circadian system in photoperiodic regulation of insect antifreeze proteins. *Journal of Experimental Zoology,* **219**, 267–70.

Horwath, K. L. and Duman, J. G. (1984). Yearly variations in the overwintering mechanisms of the cold-hardy beetle *Dendroides canadensis. Physiological Zoology,* **57**, 40–5.

Hoshikawa, K., Tsutsui, H., Honma, K. and Sakagami, S. F. (1988). Cold resistance in 4 species of beetles overwintering in the soil with notes on the overwintering strategies of some soil insects. *Applied Entomology and Zoology,* **23**, 273–81.

Howell, P. J. (1977). Recent trends in the incidence of aphid-borne viruses in Scotland. In *Proceedings of a symposium on problems of pest and disease control in northern Britain, University of Dundee, March 23–24,* (ed. R. A. Fox), p. 85.

Hughes, R. D. (1960). Induction of diapause in *Erioischia brassicae* Bouché (Dipt Anthomyiidae). *Journal of Experimental Biology,* **37**, 218–23.

Hussey, N. W. (1955). The life histories of *Megastigmus spermotrophus* Wachtl. (Hymenoptera: Chalcidoidea) and its principle parasite, with descriptions of the developmental stages. *Transactions of the Royal Entomological Society, London,* **106**, 133–51.

Ignatowicz, S. (1986). Crossing between non-diapausing and diapausing allopatric strains of the two-spotted spider mite *Tetranychus urticae* Koch (Acarina, Tetranychidae). *Polskie Pismo Entomologiczne,* **56**, 677–85.

Ingrisch, S. (1985). Effect of hibernation length on termination of diapause in European Tettingoniidae (Insecta: Orthoptera). *Oecologia,* **65**, 376–81.

Ingrisch, S. (1986). The plurennial life cycles of the European Tettigoniidae (Insecta: Orthoptera). 3. The effect of drought and the variable duration of the initial diapause. *Oecologia,* **70**, 624–30.

Ingrisch, S. (1987). Effect of photoperiod on the maternal induction of an egg diapause in the grasshopper *Chorthippus bomhalmi. Entomologia Experimentalis et Applicata,* **45**, 133–8.

Istock, C. A., Wasserman, S. S. and Zimmer, H. (1975). Ecology and evolution of the pitcher-plant mosquito. 1. Population dynamics and laboratory responses to food and population density. *Evolution,* **29**, 296–312.

Ito, K. (1988a). Effects of feeding and temperature on the hiding behaviour of *Cletus punctiger* Dallas (Heteroptera: Coreidae) in hibernacula. (In Japanese). *Japanese Journal of Applied Entomology and Zoology,* **32**, 49–54.

Ito, K. (1988b). Diapause termination in *Cletus punctiger* Dallas (Heteroptera: Coreidae) in the field. (In Japanese). *Japanese Journal of Applied Entomology and Zoology,* **32**, 63–7.

Ivens, G. W. (1988). *The UK pesticide guide.* CAB International, British Crop Protection Council.

James, B. D. and Luff, M. L. (1982). Cold-hardiness and development of eggs of *Rhopalosiphum insertum. Ecological Entomology,* **7**, 277–82.

James, D. G. (1984). Phenology of weight, moisture and energy reserves of Australian monarch butterflies *Danaus plexippus. Ecological Entomology,* **9**, 421–8.

Jensen, R. E., Savage, E. F. and Hayden, R. A. (1970). The effect of certain environmental factors on cambium temperature of peach trees. *Journal of the American Horticultural Society*, **95**, 286–91.

Johansson, J. O. (1981). Att rakna haggägg. *Lantmannen*, **102**, 15–16.

Jones, F. G. W. and Dunning, R. A. (1972). *Sugar beet pests*. Bulletin of the Ministry of Agriculture, Fisheries and Food, no. 162, HMSO, London.

Joose, E. N. G. (1966). Some observations on the biology of *Anurida maritima* (Gúerin), (Collembola). *Zeitschrift für Morphologie und Ökologie der Tiere*, **57**, 320–8.

Jørgensen, J. and Thygensen, T. (1968). Gulerodsfluen, *Psila rosae* F. *Tidsskrift for Planteval*, **72**, 1–25.

Judge, F. D. (1967). Overwintering in *Pemphigus bursarius* (L.). *Nature*, **216**, 1041–2.

Kalpage, K. S. P. and Brust, R. A. (1974). Studies on diapause and female fecundity in *Aedes atropalpus*. *Environmental Entomology*, **3**, 139–45.

Kato, Y. (1986). The prediapause copulation and its significance in the butterfly *Eurema hecabe*. *Journal of Ethology*, **4**, 81–90.

Kato, Y. and Sano, M. (1987). Role of photoperiod and temperature in seasonal morph determination of the butterfly *Eurema hecabe*. *Physiological Entomology*, **12**, 417–23.

Kaufmann, T. (1971). Hibernation in the arctic beetle, *Pterostichus brevicornis*, in Alaska. *Journal of the Kaula Entomological Society*, **44**, 87–92.

Kefuss, J. A. (1978). Influence of photoperiod on the behaviour and brood-rearing activities of honeybees in a flight room. *Journal of Apicultural Research*, **17**, 137–51.

Kidd, N. A. C. and Tozer, D. J. (1985). Distribution, survival and hatching of overwintering eggs in the large pine aphid, *Cinara pinea* (Mordv.) (Hom., Lachnidae). *Zeitschrift für Angewandte Entomologie*, **100**, 17–23.

Kimura, M. T. (1988a). Interspecific and geographical variation of diapause intensity and seasonal adaptation in the *Drosophila auraria* species complex (Diptera: Drosophilidae). *Functional Ecology*, **2**, 177–83.

Kimura, M. T. (1988b). Male mating activity and genetic aspects in imaginal diapause of *Drosophila triauraria*. *Entomologia Experimentalis et Applicata*, **47**, 81–8.

Kipyatkov, V. Y. (1974). A study of the photoperiodic reaction in the ant *Myrmica rubra* L. (Hymenoptera, Formicidae). Communication 1. Basic parameters of the reaction. *Entomological Review*, **53**, 35–41.

Kipyatkov, V. Y. (1976). A study of the photoperiodic reaction in the ant *Myrmica rubra* L. (Hymenoptera, Formicidae). Communication 5. Perception of photoperiodic information by an ant colony. *Entomological Review*, **55**, 27–34.

Kipyatkov, V. Y. (1979). The ecology of photoperiodism in the ant *Myrmica rubra* (Hymenoptera, Formicidae) 1. Seasonal changes in the photoperiodic reaction. *Entomological Review*, **58**, 10–19.

Klimetzek, D. (1972). Die Zitfolge von Ubervermehrungen nadelfressender Kiefernraupen in der Pflanz seit 1810 und die Ursachen ihres Ruckganges in neuerer Zeit. *Zeitschrift für Angewandte Entomologie*, **71**, 414–28.

Knight, J. D. and Bale, J. S. (1986). Cold hardiness and overwintering of the grain aphid *Stobion avenae*. *Ecological Entomology*, **11**, 189–97.

Knight, C. A., DeVries, A. L. and Oolman, L. D. (1984). Fish antifreeze protein and the freezing and recrystallisation of ice. *Nature*, **308**, 295–6.

Knight, J. D., Bale, J. S., Gleave, H. and Stevenson, R. A. (1986a). A multi-channel micro-bead thermometer for the detection of low level thermal energy changes in small invertebrates. *Cryo-Letters*, **7**, 299–304.

Knight, J. D., Bale, J. S., Franks, F., Mathias, S. F. and Baust, J. G. (1986b). Insect cold hardiness: supercooling points and pre-freeze mortality. *Cryo-Letters*, **7**, 194–203.

Kobayashi, S., Ide, M. and Higashi, O. (1988). Overwintering of the rice water weevil, *Lissorhoptrus oryzophilus* Kuschel (Coleoptera: Curculionidae), under snow and freezing conditions at a high elevation. (In Japanese). *Japanese Journal of Applied Entomology and Zoology*, **32**, 79–82.

Komazaki, S. (1983). Overwintering of the spirea aphid, *Aphis citricola* Van der Goot (Homoptera: Aphididae) on citrus and spirea plants. *Applied Entomology and Zoology*, **18**, 301–7.

Kotaki, T. and Yagi, S. (1987). Relationship between diapause development and coloration change in brown-winged green bug, *Plautia stali* Scott (Heteroptera: Pentatomidae). (In Japanese). *Japanese Journal of Applied Entomology and Zoology*, **31**, 235–90.

Koveos, D. S. and Tzanakakis, A. (1989). Influence of photoperiod, temperature and host plant in the production of diapause eggs in *Petrobia* (*Tetranychina*) *harti* (Acari: Tetranychidae). *Experimental and Applied Acarology*, **6**, 327–42.

Krog, J. O., Zachariassen, K. E., Larsen, B. and Smidsrod, O. (1979). Thermal buffering in Afro-Alpine plants due to nucleating agent-induced water freezing. *Nature*, **282**, 300–1.

Krombein, K. V., Hurd, P. D. Jr, Smith, D. R. and Busks, B. D. (1979). *Catalogue of Hymenoptera in America North of Mexico*. Smithsonian Institute Press, Washington DC.

Kudo, Sin-ichi and Kurihara, Morihisa (1988). Seasonal occurrence of egg diapause in the rice leaf bug, *Trigonotylus coelestialium* Kirkcaldy (Hemiptera: Miridae). *Applied Entomology and Zoology*, **23**, 365–6.

Kukal, O. and Kevan, P. G. (1987). The influence of parasitism on the life history of a high Arctic insect, *Gynaephora groenlandica* (Wocke) (Lepidoptera: Lymantriidae). *Canadian Journal of Ecology*, **65**, 156–63.

Kulman, H. M. (1967). Within-tree distribution and winter mortality of eggs of the woolly pine needle aphid, *Schizolachnus piniradiatae*. *Annals of the Entomological Society of America*, **60**, 384–6.

Kurahashi, H. and Ohtaki, T. (1989). Geographical variation in the incidence of pupal diapause in Asian and Oceanian species of the flesh fly *Boettcherisca* (Diptera: Sarcophagidae). *Physiological Entomology*, **14**, 291–8.

Kurppa, S. (1986). Erfarenheter av bladlusaret 1985 i Finland. *Vaxtskyddsräpporter Jordbruk*, **39**, 125–35.

Kurppa, S. (1989). Predicting outbreaks of *Rhopalosiphum padi* in Finland. *Annals Agriculturae Fenniae*, **28**, 333–48.

Kurtak, D. (1974). Overwintering of *Simulium pictipes* Hagen (Diptera: Simuliidae) as eggs. *Journal of Medical Entomology*, **11**, 383–4.

Kuznetsova, I. A. and Tyshehenko, V. P. (1979). The part played by photoperiod and temperature in regulating cessation of diapause in *Drosophila transversa* FU and *D. pholerata* Mg (Diptera, Drosophilidae). *Entomological Review*, **58**, 8–15.

Labanowski, G. (1981). Spring emergence of the codling moth, *Laspeyresia pomonella* (L.) and the possibility of forecasting it in central Poland. *Ekologia Polska*, **29**, 535–44.

Laing, J. E. and Heraty, J. M. (1984). The use of day degrees to predict emergence of the apple maggot *Rhagalatis pomonella* (Diptera: Tephritidae), in Ontario. *Canadian Entomologist*, **116**, 1223–9.

Lamb, R. J., Turnock, W. J. and Hayhoe, H. N. (1985). Winter survival and outbreaks of bertha armyworm, *Mamestra configurata* (Lepidoptera: Noctuidae), on Canola. *Canadian Entomologist*, **117**, 727–36.

Leather, S. R. (1980a). Egg survival in the bird cherry-oat aphid, *Rhopalosiphum padi*. *Entomologia Experimentalis et Applicata*, **27**, 96–7.

Leather, S. R. (1980b). Aspects of the ecology of the bird cherry-oat aphid, *Rhopalosiphum padi L.* Unpublished PhD Thesis, University of East Anglia, Norwich.

Leather, S. R. (1981). Factors affecting egg survival in the bird cherry-oat aphid, *Rhopalosiphum padi*. *Entomologia Experimentalis et Applicata*, **30**, 197–9.

Leather, S. R. (1982). Viljakasvien kirvat, niiden vahingollisuus, torjunta ja ennustaminen. *Kaytannon Maamies*, **1982**, 28–30.

Leather, S. R. (1983). Forecasting aphid outbreaks using winter egg counts: an assessment of its feasibility and an example of its applications in Finland. *Zeitschrift für angewandte Entomologie*, **96**, 282–7.

Leather, S. R. (1984). Factors affecting pupal survival and eclosion in the pine beauty moth, *Panolis flammea* (D & S). *Oecologia*, **63**, 75–9.

Leather, S. R. (1986). Host monitoring by aphid migrants: do gynoparae maximise offspring fitness? *Oecologia (Berlin)*, **68**, 367–9.

Leather, S. R. (1990). The role of host quality, natural enemies, competition and weather in the population regulation of the bird cherry aphid. In *Population dynamics of forest insects*, (ed. A. D. Watt, S. R. Leather, N. A. C. Kidd and M. D. Hunter), pp. 35–44. Intercept Books.

Leather, S. R. and Brotherton, C. M. (1987). Defensive responses of the pine beauty moth, *Panolis flammea* (D & S) (Lepidoptera: Noctuidae). *Entomologist's Gazette*, **38**, 19–24.

Leather, S. R. and Lehti, J. P. (1981). Abundance and survival of eggs of the bird cherry-oat aphid, *Rhopalosiphum padi* in southern Finland. *Annales Entomologici Fennici*, **47**, 125–30.

Leather, S. R. and Lehti, J. P. (1982). Field studies on the factors affecting the population dynamics of the bird cherry-oat aphid, *Rhopalosiphum padi* (L.) in Finland. *Annales Agriculturae Fenniae*, **21**, 20–31.

Leather, S. R. and Walters, K. F. A. (1984). Spring migration of cereal aphids. *Zeitschrift für angewandte Entomologie*, **97**, 431–7.

Leather, S. R., Watt, A. D. and Barbour, D. A. (1985). The effect of host-plant and delayed mating on the fecundity and lifespan of the pine beauty moth, *Panolis flammea* (Dennis and Schiffermuller) (Lepidoptera: Noctuidae): their influence on population dynamics and relevance to pest management. *Bulletin of Entomological Research*, **75**, 641–51.

Leather, S. R., Walters, K. F. A. and Dixon, A. F. G. (1989). Factors determining the pest status of the bird cherry-oat aphid, *Rhopalosiphum padi* (L.) (Hemiptera: Aphididae) in Europe: a study and review. *Bulletin of Entomological Research*, **79**, 345–60.

Lee, R. E. and Denlinger (1985). Cold tolerance in diapausing and non-diapausing stages of the flesh fly *Sarcophage crassipalpis*. *Physiological Entomology*, **10**, 309–15.

Lee, R. E., Ring, R. A. and Baust, J. G. (1986). Low temperature tolerance in insects and other terrestrial arthropods: Bibliography II. *Cryo-Letters*, 7, 113–26.

Lees, A. D. (1955). *The physiology of diapause in arthropods.* Cambridge University Press, London.

Lees, A. D. (1968). Photoperiodism in insects. In *Photophysiology* (ed. A. C. Giese), pp. 47–137. Academic Press, London.

Lees, A. D. (1973). Photoperiodic time measurement in the aphid *Megoura viciae*. *Journal of Insect Physiology*, 19, 2279–316.

Lees, A. D. (1989). The photoperiodic responses and phenology of an English strain of the pea aphid *Acyrthosiphon pisum*. *Ecological Entomology*, 14, 69–78.

Lees, E. and Tilley, R. J. D. (1980). Influence of photoperiod and temperature on larval development in *Pararge aegeria* (L.). (Lepidoptera: Satyridae). *Entomologist's Gazette*, 31, 3–6.

Lefevere, K. S. and DeKort, C. A. (1989). Adult diapause in the Colorado Beetle, *Leptinotarsa decemlineata*: effects of external factors on maintenance, termination and postdiapause development. *Physiological Entomology*, 14, 299–308.

Leonard, D. E. (1972). Survival in a gypsy moth population exposed to low winter temperatures. *Environmental Entomology*, 1, 549–54.

Leonard, D. E. (1974). Ecology and control of the gypsy moth. *Annual Review of Entomology*, 19, 197–229.

Levins, R. (1969). Dormancy as an adaptive strategy. *Symposium of the Society of Experimental Biology*, 23, 1–10.

Levitt, J. A. (1962). A sulfhydryl-disulphide hypothesis of frost injury and resistance in plants. *Journal of Theoretical Biology*, 3, 355–91.

Levitt, J. A. (1980). *Responses of plants to environmental stresses. Physiological Ecology Volume 1. Chilling, freezing and high temperature stresses.* Academic Press, London.

Lewis, T. and Navas, D. E. (1962). Thysanopteran populations overwintering in hedge bottoms, grass litter and bark. *Annals of Applied Biology*, 50, 299–311.

Lovelock, J. E. (1953). The mechanism of the protective action of glycerol against haemolysis by freezing and thawing. *Biochimica et Biophysica Acta*, 11, 28–36.

Lozina-Lozinskii, L. K. (1974). *Studies in cryobiology*, pp. 69–84. John Wiley and Sons, New York.

Lumme, J. and Keränen, L. (1978). Photoperiodic diapause in *Drosophila lummei* Hackman is controlled by an X-chromosomal factor. *Hereditas*, 89, 221–62.

Lumme, J. and Pohjola, L. (1980). Selection against photoperiodic diapause started from monohybrid crosses in *Drosophila littoralis*. *Hereditas*, 92, 377–8.

Luyet, B. J. (1960). On various phase transitions occurring in aqueous solutions at low temperature. *Annals of the New York Academy of Sciences*, 85, 549–69.

Mackay, J. R. and Mackay, D. K. (1974). Snow cover and ground temperatures, Camy Island, NWT. *Arctic*, 27, 287–97.

MacKenzie, A. P. (1977). Non-equilibrium freezing behaviour of aqueous systems. *Philosophical Transactions of the Royal Society, Series B*, 278, 167–89.

MacLean, S. F. (1973). Life cycle and growth energetics of the arctic crane fly. *Pedicia nannai antennata. Oikos*, 24, 436–43.

MacLeod, E. G. (1967). Experimental induction and elimination of adult diapause and autumnal coloration in *Chrysopa carnea* (Neuroptera). *Journal of Insect Physiology*, **13**, 1343–9.

Madrid, F. J. and Stewart, R. K. (1981). Ecological significance of cold hardiness and winter mortality of eggs of the gypsy moth *Lymantria dispar* L., in Quebec. *Environmental Entomology*, **10**, 586–9.

Mail, G. A. (1930). Winter soil temperatures and their relation to subterranean insect survival. *Journal of Agricultural Research*, **41**, 571–92.

Maltais, J., Régnière, J., Cloutier, C., Hébert, C. and Perry, D. F. (1989). Seasonal biology of *Meteorus trachynotus* Vier. (Hymenoptera: Braconidae) and of its overwintering host *Choristoneura rosaceana* (Harr.) (Lepidoptera: Tortricidae). *Canadian Entomologist*, **121**, 745–56.

Mansingh, A. (1971). Physiological classification of dormancies in insects. *Canadian Entomologist*, **103**, 983–1009.

Mansingh, A. and Smallman, B. N. (1972). Variation in polyhydric alcohol in relation to diapause and cold-hardiness in the larvae of *Isia isabella*. *Journal of Insect Physiology*, **18**, 1565–71.

Mansingh, A. and Steele, R. W. (1973). Studies on insect dormancy. I. Physiology of hibernation in the larvae of the black fly *Prosimuluim mysticum* Peterson. *Canadian Journal of Zoology*, **51**, 611–18.

Masaki, S. (1956). The effect of temperature on the termination of diapause in the egg of *Lymantria dispar* Linneé. *Japanese Journal of Applied Zoology*, **21**, 148–57.

Masaki, S. (1978). Seasonal and latitudinal adaptations in the life cycles of crickets. In *Evolution of insect migration and diapause*, (ed H. Dingle), pp. 72–100. Springer-Verlag, New York.

Masaki, S., Ando, Y. and Watanabe, A. (1979). High temperature and diapause termination in the eggs of *Teleogryllus commodus* (Orthoptera: Gryllidae). *Kontyû*, **47**, 493–504.

Maslennikova, V. A. (1958). The conditions determining diapause in the parasitic Hymenopteran *Apanteles glomeratus* L. (Hymenoptera, Braconidae) and *Pteromalus pupanum* L. (Hymenoptera, Chalcididae). *Entomological Review*, **37**, 466–72.

Mason, R. R. and Wickman, B. E. (1988). The Douglas-fir tussock moth in the interior Pacific Northwest. In *Dynamics of forest insect populations: patterns causes and implications*, (ed. A. A. Berryman), pp. 179–209. Plenum Press, New York.

Mason, R. R., Beckwith, R. C. and Paul, H. G. (1977). Fecundity reduction during collapse of a Douglas-fir tussock moth outbreak in northeast Oregon. *Environmental Entomology*, **6**, 623–6.

Mason, R. R., Torgersen, T. R., Wickman, B. E. and Paul, H. G. (1983). Natural regulation of a Douglas-fir tussock moth (Lepidoptera: Lymantriidae) population in the Sierra Nevada. *Environmental Entomology*, **12**, 587–94.

Matthew, D. L. (1961). Face fly overwintering in Indiana homes. *Proceedings of the North Central Branch, American Association of Economic Entomologists*, **16**, 111–12.

May, M. L. (1979). Insect thermoregulation. *Annual Review of Entomology*, **24**, 313–49.

McLeod, D. G. R., Ritchat, C. and Nagai, T. (1979). Occurrence of a two generation strain of the European corn borer, *Ostrinia nubilalis* (Lepidoptera: Pyralidae), in Quebec. *Canadian Entomologist*, **111**, 2330–6.

McLeod, D. G. R., Whistlecraft, J. W. and Harris, C. R. (1985). An improved rearing procedure for the carrot rust fly (Diptera: Psilidae) with observations on life history and conditions controlling diapause induction and termination. *Canadian Entomologist*, **117**, 1017–24.

McLeod, P. (1987). Effect of low temperature on *Myzus persicae* (Homoptera: Aphididae) on overwintering spinach. *Environmental Entomology*, **16**, 796–801.

McNeil, J. N. and Rabb, R. L. (1973). Physical and physiological factors in diapause initiation of two hyperparasites of the tobacco hornworm, *Manduca sexta*. *Journal of Insect Physiology*, **19**, 2107–18.

Meinke, L. J. and Gould, F. (1987). Thermoregulation by *Diabrotica undecimpunctata howardi* and potential effects on overwintering biology. *Entomologia Experimentalis et Applicata*, **45**, 115–22.

Mellini, E. (1983). L'ipotesi della dominazione armonale, esercita dogli aspiti sun parassitoide, alla luce delle recenti scoperte nella endocrinologia deglie insetti. *Bolletino dell'Istituto di Entomologia della Università degli Studi di Bologna*, **38**, 135–66.

Meryman, H. T. (1966). Review of biological freezing. In *Cryobiology*, (ed. H. T. Meryman), pp. 1–106. Academic Press, London.

Meryman, H. T. (1970). The exceeding of a minimum tolerance cell volume in hypertonic suspension as a cause of freezing injury. In *The frozen cell*, (ed. G. Wolstenholme and M. O'Connor), pp. 51–68. Churchill, London.

Mikkola, K. (1980). Miten perhoset selviävät pakkasista. *Suomen Luonto*, **39**, 12–15.

Miller, L. K. (1969). Freezing tolerance in an adult insect. *Science*, **116**, 105–6.

Miller, L. K. (1982). Cold-hardiness strategies of some adult and immature insects overwintering in interior Alaska. *Comparative Biochemistry and Physiology*, **73**, 595–604.

Ministry of Agriculture, Fisheries and Food. (1979). *Leatherjackets*. Ministry of Agriculture, Fisheries and Food, Advisory Leaflet **179**. London.

Ministry of Agriculture, Fisheries and Food. (1982). *Wheat bulb fly*. Ministry of Agriculture, Fisheries and Food, Advisory Leaflet **177**. London.

Ministry of Agriculture, Fisheries and Food. (1983). *Cutworms*. Ministry of Agriculture, Fisheries and Food, Advisory Leaflet **225**. London.

Missonnier, J. (1963). Etude écologique du développement nymphal de deux Diptères Muscides phytophages: *Pegomyia betae* Curtis et *Chortophila brassicae* Bouche. *Annales des Epiphyties*, **14**, 293–310.

Moore, S. D. (1989). Regulation of diapause by an insect parasitoid. *Ecological Entomology*, **14**, 93–8.

Monro, H. A. V. (1935). Observations on the habits of an introduced pine sawfly, *Diprion simile* Htg. *Canadian Entomologist*, **67**, 137–40.

Montgomery, M. E. and Wallner, W. E. (1988). The gypsy moth – a westward migrant. In *Dynamics of forest insect populations: patterns, causes implications*, (ed. A. A. Berryman), pp. 353–75. Plenum Press, New York.

Mopper, S., Faeth, S. H., Boecklen, W. J. and Simberloff, D. S. (1985). Host-specific variation in leaf miner population dynamics: effects on density, natural enemies and behaviour of *Stilbosis quadricustatella* (Lepidoptera: Cosmopterigidae). *Ecological Entomology*, **9**, 169–77.

Morrisey, R. E. and Baust, J. G. (1976). The ontogeny of cold tolerance in the gall fly *Eurosta solidaginis*. *Journal of Insect Physiology*, **22**, 431–7.

Mukerji, M. K. and Braun, M. P. (1988). Effect of low temperatures on mortality of grasshopper eggs (*Orthoptera: Acrididae*). *Canadian Entomologist*, **120**, 1147–8.

Müller, H. J. (1962). Über die Induktion der Diapause und der Ausbildung der Saisonformen bei *Aleurochiton complanatus* (Baerensprung). *Zeitschrift für Morphologie und Ökologie der Tiere*, **51**, 575–610.

Nagell, B. (1977). Phototactic and thermotactic responses facilitating survival of *Cloeon dipterum* (Ephemeroptera) larvae under winter anoxia. *Oikos*, **29**, 342–7.

Naton, E. (1966). Voraussetzungen für eine Laborzucht der echten Möhrenfliege, *Psila rosae* Fabr. *Anzeiger für Schaedlingskunde Pflanzenschutz, Umweltschutz*, **39**, 85–9.

Nealis, V. (1985). Diapause and the seasonal ecology of the introduced parasite *Cotesia* (*Apanteles*) *rubecula* (Hymenoptera: Braconidae). *Canadian Entomologist*, **117**, 333–42.

Nechols, J. R. (1987). Voltinism, seasonal reproduction and diapause in the squash bug (Heteroptera: Coreidae) in Kansas. *Environmental Entomology*, **16**, 269–73.

Nechols, J. R. (1988). Photoperiodic responses of the squash bug (Heteroptera: Coreidae): diapause induction and maintenance. *Environmental Entomology*, **17**, 427–31.

Nechols, J. R., Tanker, M. J. and Helgeren, R. G. (1980). Environmental control of diapause and postdiapause development in *Tetrastichus julis* (Hymenoptera: Eulophidae), a parasite of the cereal leaf beetle, *Oulema melanoplus* (Coleoptera: Chrysomelidae). *Canadian Entomologist*, **112**, 1277–84.

Neilson, W. T. A. (1962). Effects of temperature on development of overwintering pupae of the apple maggot, *Rhagoletis pomonella* (Walsh). *Canadian Entomologist*, **94**, 924–8.

Neudecker, C. (1974). Das Präferenzverhalten von *Agonum assimile* Payk. (Carab. Coleopt.) in Temperatur-, Feuchtigkeits- und Helligkeits-gradienteu. *Zoologisches Jahrbuch Arbeit Systematik Ökologie Geographie Tiere*, **101**, 609–27.

Neudecker, C. and Thiele, H. U. (1974). Die jahreszeitliche Synchronisation der Gonadenreifung bei *Agonum assimile* Payk (Coleopt Carab) durch Temperatur und Photoperiode. *Oecologia*, **17**, 141–57.

Nothnagle, P. J. and Schultz, J. C. (1987). What is a forest pest? In *Insect outbreaks*, (ed. P. Barbosa and J. C. Schultz), pp. 59–80. Academic Press, New York.

Numate, N. and Hidaka, T. (1982). Photoperiodic control of adult diapause in the bean bug, *Riptortus clavatus* Thunberg (Heteroptera: Coreidae). 1. Reversible induction and termination of diapause. *Applied Entomology and Zoology*, **17**, 530–8.

Nylin, S. (1989). Effects of changing photoperiods in the life cycle regulation of the common butterfly, *Polygonia c-album* (Nymphalidae). *Ecological Entomology*, **14**, 209–18.

Oakley, J. N. and Uncles, J. J. (1977). Use of oviposition traps to estimate numbers of wheat bulb fly, *Leptohylemia coarctata* eggs. *Annals of Applied Biology*, **85**, 407–9.

Obrycki, J. J. and Tauber, M. J. (1979). Seasonal synchrony of the parasite *Perilitus coccinellae* and its host, *Coleomegilla maculata*. *Environmental Entomology*, **8**, 400–5.

Obrycki, J. J., Tauber, M. J., Tauber, C. A. and Gallands, B. (1983).

Environmental control of the seasonal life cycle of *Adalia bipunctata* (Coleoptera: Coccinellidae). *Environmental Entomology*, **12**, 416–21.

O'Doherty, R. (1986). Cold hardiness of laboratory-maintained and seasonally-collected populations of the black bean aphid *Aphis fabae* Scopoli (Hemiptera: Aphididae). *Bulletin of Entomological Research*, **76**, 367–74.

O'Doherty, R. and Bale, J. S. (1985). Factors affecting the cold hardiness of the peach-potato aphid *Myzus persicae*. *Annals of Applied Biology*, **106**, 219–28.

Oku, T. (1982). Overwintering of eggs in the Siberian cutworm *Euxoa sibirica* Boisduval (Lepidoptera: Noctuidae). *Applied Entomology Zoology*, **17**, 244–52.

Parish, W. E. G. and Bale, J. S. (1990). The effect of feeding and gut contents on supercooling in larvae of *Pieris brassicae*. *Cryo-Letters*, **11**, 67–74.

Park, O. (1930). Studies in the ecology of forest Coleoptera. Seral and seasonal succession of Coleoptera in the Chicago area, with observations on certain phases of hibernation and aggregation. *Annals of the Entomological Society of America*, **23**, 57–80.

Parrish, D. S. and Davis, D. W. (1978). Inhibition of diapause in *Bathyplectes curculionis*, a parasite of the alfalfa weevil. *Annals of the Entomological Society of America*, **71**, 103–7.

Parry, W. H. (1978). Studies on the factors affecting the population levels of *Adelges cooleyi* (Gillette) on Douglas fir. I. Sistentes on mature needles. *Zeitschrift für angewandte Entomologie*, **85**, 365–78.

Parry, W. H. (1980). Studies on the factors affecting the population levels of *Adelges cooleyi* (Gillette) on Douglas fir. III. Low temperature mortality. *Zeitschrift für angewandte Entomologie*, **90**, 133–41.

Parry, W. H. (1981). Weevils (Coleoptera: Curculionidae) on foliage of Sitka spruce in north east Scotland. *Scottish Forestry*, **35**, 96–101.

Parry, W. H. (1983). The overwintering of adult *Strophosomus melanogrammus* (Coleoptera: Curculionidae) in forest soils. In *Proceedings of the VIII International Colloquium on Soil Zoology*, (ed. P. Lebrun, H. M. Andre, A. De Medts, C. Gregoire-wibo and G. Wauthy), pp. 433–9.

Patterson, J. L. and Duman, J. G. (1978). The role of the thermal hysteresis factor in *Tenebrio molitor* larvae. *Journal of Experimental Biology*, **74**, 37–45.

Payne, N. M. (1927a). Freezing and survival of insects at low temperatures. *Journal of Morphology*, **43**, 521–46.

Payne, N. M. (1927b). Measures of insect cold-hardiness. *Biological Bulletin*, **52**, 449–57.

Pease, W. R. (1962). Factors causing seasonal forms in *Ascia monuste* (Lepidoptera). *Science*, **137**, 987–8.

Pener, M. P. (1974). Neurosecretory and corpus allatum controlled effects on male sexual behaviour in acridids. In *Experimental analysis of insect behaviour*, (ed. L. Barton Browne), pp. 264–77. Springer-Verlag, Berlin.

Pener, M. P. and Broza, M. (1971). The effect of implanted, active corpora allata on reproductive diapause in adult females of the grasshopper *Oedipoda miniata*. *Entomologia Experimentalis et Applicata*, **14**, 190–202.

Pener, M. P. and Orshan, L. (1980). Reversible reproductive diapause and intermediate states between diapause and full reproductive activity in male *Oedipoda miniata* grasshoppers. *Physiological Entomology*, **5**, 417–26.

Peterson, A. (1917). Studies on the morphology and susceptibility of the eggs of

Aphis avenae Fab., *Aphis pomi* De Geer and *Aphis sorbi* Kalt. *Journal of Economic Entomology,* **10**, 555–60.

Peterson, A. (1920). Some studies on the influence of environmental factors on the hatching of the eggs of *Aphis avenae* Fabricus and *Aphis pomi* De Geer. *Annals of the Entomological Society of America,* **13**, 391–401.

Peterson, D. M. and Hamner, W. M. (1968). Photoperiodic control of diapause in the codling moth. *Journal of Insect Physiology,* **14**, 519–28.

Pienkowski, R. L. (1976). Behaviour of the adult alfalfa weevil in diapause. *Annals of the Entomological Society of America,* **69**, 155–7.

Plantevin, G., Grenier, S., Richard, G. and Norden, C. (1986). Larval development, developmental arrest and hormonal levels in the couple *Galleria mellonella* (Lepidoptera-Pyralidae) – *Pseudoperichaeta nigratinenta* (Diptera-Tachinidae). *Archives of Insect Biochemistry,* **3**, 457–69.

Pogue, M. G. (1985). Parasite complex of *Archips argyrospilus, Choristoneura rosaceana* (Lepidoptera: Tortricidae) and *Anacampsis innocuella* (Lepidoptera: Gelechiidae) in Wyoming shelterbelts. *Entomological News,* **96**, 83–6.

Pointing, P. J. (1963). The biology and behaviour of the European pine shoot moth, *Rhyacionia bioliana* (Schiff.) in southern Ontario II. Egg, larva and pupa. *Canadian Entomologist,* **95**, 844–63.

Pouvreau, A. (1970). Données écologiques sur l'hibernation contrôleé des reines de bourdons (Hymenoptera, Apoidea, Bombinae, *Bombus,* Leitr.) *Apidologie,* **1**, 73–95.

Powell, W. (1976). Supercooling temperature distribution curves as possible indicators of aphid food quality. *Journal of Insect Physiology,* **22**, 595–9.

Price, J. R., Slosser, J. E. and Puterka, G. J. (1985). Factors affecting survival of boll weevils (Coleoptera: Curculionidae) in winter habitat in Texas rolling plains. *Southwestern Entomologist,* **10**, 1–6.

Proctor, N. S. (1976). Mass hibernation site for *Nymphalis vau-album* (Nymphalidae). *Journal of the Lepidopterist's Society,* **30**, 126.

Pruitt, W. O. (1979). Some ecological aspects of snow. In *Ecology of the subarctic regions,* pp. 83–99. UNESCO, Paris.

Pryor, M. E. (1962). Some environmental features of Hallett station, Antarctica, with special reference to soil arthropods. *Pacific Insects,* **4**, 681–728.

Pullin, A. S. (1986). Effect of photoperiod and temperature on the life-cycle of different populations of the peacock butterfly *Inachis io. Entomologia Experimentalis et Applicata,* **41**, 237–42.

Pullin, A. S. (1987). Adult feeding time, lipid accumulation, and overwintering in *Aglais urticae* and *Inachis io* (Lepidoptera: Nymphalidae). *Journal of Zoology,* **211**, 631–42.

Pullin, A. S. and Bale, J. S. (1988). Cause and effects of pre-freeze mortality in aphids. *Cryo-Letters,* **9**, 101–13.

Quiring, D. T. and Timmins, P. R. (1988). Predation by American crows reduces overwintering European corn borer population in southwestern Ontario. *Canadian Journal of Zoology,* **66**, 2143–5.

Ramadhane, A., Grenier, S. and Plantevin, G. (1987). Physiological interactions and development synchronizations between non-diapausing *Ostrinia nubilalis* larvae and the tachinid parasitoid *Pseudoperichaeta nigrolineata. Entomologia Experimentalis et Applicata,* **45**, 157–65.

Ramadhane, A., Grenier, S. and Plantevin, G. (1988). Photoperiod, temperature and ecdysteroid influences on physiological interactions between diapausing *Ostrinia nubilalis* larvae and the tachinid parasitoid

Pseudoperichaeta nigrolineata. Entomologia Experimentatis et Applicata, **48**, 275–82.

Ramsay, J. (1964). The rectal complex of the mealworm *Tenebrio molitor* L. (Coleoptera, Tenebrionidae). *Philosophical Transactions of the Royal Society B*, **248**, 279–314.

Rasnitsyn, A. P. (1966). Overwintering of Ichneumon-flies (Hymenoptera: Ichneumonidae). *Entomological Review*, **43**, 24–6.

Rautapää, J. (1976). Population dynamics of cereal aphids and method of predicting population trends. *Annales Agriculturae Fenniae*, **15**, 272–93.

Raymond, J. A. and DeVries, A. L. (1977). Absorption inhibition as a mechanism of freezing resistance in polar fishes. *Proceedings of the National Academy of Science USA*, **74**, 2589–93.

Read, D. C. (1969). Rearing the cabbage maggot with and without diapause. *Canadian Entomologist*, **101**, 725–37.

Richey, S. J., Hainze, J. H. and Scriber, T. M. (1987). Evidence of a sex-linked diapause response in *Papilio glaucus* subspecies and their hybrids. *Physiological Entomology*, **12**, 181–4.

Rickards, J., Kelleher, M. J. and Storey, K. B. (1987). Strategies of freeze avoidance in larvae of the goldenrod gall moth *Epiblema scudderiana*: winter profiles of a natural population. *Journal of Insect Physiology*, **33**, 443–50.

Riedl, H. (1983). Analysis of codling moth phenology in relation to latitude, climate, and food availability. In *Diapause and life cycle strategies in insects*, (eds V. K. Brown and I. Hodek), pp. 233–52. Dr W. Junk Publishers, The Hague.

Riedl, H. and Croft, B. A. (1978). The effects of photoperiod and effective temperatures on the seasonal phenology of the codling moth (Lepidoptera: Tortricidae). *Canadian Entomologist*, **110**, 455–70.

Riegert, P. W. (1967). Association of subzero temperatures, snow cover and winter mortality of grasshopper eggs in Saskatchewan. *Canadian Entomologist*, **99**, 1000–3.

Riegert, P. W. (1968). A history of grasshopper abundance surveys and forecasts of outbreaks in Saskatchewan. *Memoirs of the Entomological Society of Canada*, **52**, 5–99.

Rieux, R. and D'Arcier, F. F. (1990). Polymorphisme saisonnier des populations naturelles des adultes de *Psylla pyri* (L.) (Hom., Psyllidae). *Journal of Applied Entomology*, **109(2)**, 120–35.

Riihimaa, A. (1984). The inheritance of the facultative diapause in *Chymomyza costata. Hereditas*, **101**, 283.

Riihimaa, A. and Kimura, M. T. (1988). A mutant strain of *Chymomyza costata* (Diptera: Drosophilidae) insensitive to diapause-inducing action of photoperiod. *Physiological Entomology*, **13**, 441–5.

Ring, R. A. (1977). Cold-hardiness of the bark beetle *Scolytus ratzeburgi* Jans. (Col Scolytidae). *Norwegian Journal of Entomology*, **24**, 125–36.

Ring, R. A. (1980). Insects and their cells. In *Low temperature preservation in medicine and biology*, (ed. M. J. Ashwood-Smith and J. Farrant), pp. 187–217. Pitman Medical, London.

Ring, R. A. (1982). Freezing-tolerant insects with low supercooling points. *Comparative Biochemistry and Physiology*, **73A**, 605–12.

Ring, R. A. and Tesar, D. (1980). Cold hardiness of the Arctic beetle, *Pytho americanus* Kirby, Coleoptera, Pythidae (Salpingidae). *Journal of Insect Physiology*, **26**, 763–74.

Ring, R. A. and Tesar, D. (1981). Adaptations to cold in Canadian arctic insects. *Cryobiology*, **18**, 199–211.

Rockey, S. J., Hainze, J. H. and Scriber, J. M. (1987). Evidence of a sex-linked diapause response in *Papilio glaucus* subspecies and their hybrids. *Physiological Entomology*, **12**, 181–4.

Rogerson, J. P. (1947). The oat-bird cherry aphid *Rhopalosiphum padi* (L.) and comparison with *R. crataegellum* Theo. (Hemiptera, Aphididae). *Bulletin of Entomological Research*, **38**, 157–76.

Rojas, R. R., Lee, R. E., Luu, T. A. and Baust, J. G. (1983). Temperature dependence-independence of antifreeze turnover in *Eurosta solidaginis* (Fitch). *Journal of Insect Physiology*, **29**, 865–9.

Rojas, R. R., Lee, R. E. and Baust, J. G. (1986). Relationship of environmental water content to glycerol accumulation in the freezing tolerant larvae of *Eurosta solidaginis* (Fitch). *Cryo-Letters*, **7**, 235–45.

Salt, R. W. (1936). Studies on the freezing process in insects. *Technical Bulletin Minnesota Agricultural Experimental Station*, **116**, 1–41.

Salt, R. W. (1950). Time as a factor in the freezing of undercooled insects. *Canadian Journal of Research D*, **28**, 285–91.

Salt, R. W. (1953). The influence of food on cold-hardiness of insects. *Canadian Entomologist*, **85**, 261–9.

Salt, R. W. (1956). Influence of moisture content and temperature on cold-hardiness in hibernating insects. *Canadian Journal of Zoology*, **34**, 283–94.

Salt, R. W. (1957). Natural occurrence of glycerol in insects and its relation to their ability to survive freezing. *Canadian Entomologist*, **89**, 491–4.

Salt, R. W. (1958). Application of nucleation theory to the freezing of supercooled insects. *Journal of Insect Physiology*, **2**, 178–88.

Salt, R. W. (1959). Role of glycerol in the cold-hardening of *Bracon cephi* (Gahan). *Canadian Journal of Zoology*, **37**, 59–69.

Salt, R. W. (1961). Principles of insect cold-hardiness. *Annual Review of Entomology*, **6**, 55–74.

Salt, R. W. (1962). Intracellular freezing in insects. *Nature*, **193**, 1207–8.

Salt, R. W. (1966a). Effect of cooling rate on the freezing temperatures of supercooled insects. *Canadian Journal of Zoology*, **44**, 655–9.

Salt, R. W. (1966b). Factors influencing nucleation in supercooled insects. *Canadian Journal of Zoology*, **44**, 117–33.

Salt, R. W. (1966c). Relation between time of freezing and temperature in supercooled larvae of *Cephus cinctus* Novt. *Canadian Journal of Zoology*, **44**, 947–52.

Salt, R. W. and James, H. G. (1947). Low temperature as a factor in the mortality of eggs of *Mantis religiosa* L. *Canadian Entomologist*, **79**, 33–7.

Saunders, D. S. (1965). Larval diapause of maternal origin: induction of diapause in *Nasonia vitripennis* (Walk) (Hymenoptera: Pteromalidae). *Journal of Experimental Biology*, **42**, 495–508.

Saunders, D. S. (1973). Thermoperiodic control of diapause in an insect: theory of internal coincidence. *Science*, **181**, 358–60.

Saunders, D. S. (1976). *Insect clocks*. Pergamon Press, Oxford.

Saunders, D. S. (1980). Some effects of constant temperature and photoperiod

on the diapause response of the flesh fly, *Sarcophaga argyrostoma*.
Physiological Entomology, **5**, 191–8.

Saunders, D. S. (1982). *Insect clocks*, 2nd edn. Pergamon Press, Oxford.

Saunders, D. S., Sutton, D. and Jarvis, R. A. (1970). The effect of host species
on diapause induction in *Nasonia vitripennis*. *Journal of Insect Physiology*,
16, 405–16.

Sawyer, A. J. and Haynes, D. L. (1985). Simulating the spatio-temporal
dynamics of the cereal leaf beetle in a regional crop system. *Ecological
Modelling*, **30**, 83–104.

Saynor, M. and Seal, K. (1976). Observations on the colonisation and
egg-laying behaviour of *Aphis fabae* Scop. on various species of *Euonymus*.
Plant Pathology, **25**, 209–10.

Schaller, F. (1968). Action de la température sur la diapause et le
développement de l'embryon d'*Aeshna mixta* (Odonata). *Journal of Insect
Physiology*, **14**, 1477–83.

Schmieder, R. G. (1933). The polymorphic forms of *Melittobia chalybii*
Ashwood and the determining factors involved in their production
(Hymenoptera: Chalcidoidea, Eulophidae). *Biological Bulletin*, **65**, 338–54.

Schmieder, R. G. (1939). The significance of the two types of larvae in
Sphecophaga burra (Cresson) and the factors conditioning them
(Hymenoptera: Ichneumonidae). *Entomological News*, **50**, 125–31.

Schopf, A. (1980). Zur diapause des Puppenparasiten *Pimpla turionellae* L.
(Hym., Ichneumondiae). *Zoologischer Jahresbericht Systematik*, **107**, 537–67.

Schwerdtfeger, F. (1934). Studien über den Massenwechsel einiger
Forstschadlinge III Untersuchungen über die Mortalität der Forleule
(*Panolis flammea* Schiff) im Krisenjahr einer Epidemie. *Mitt. Forstw.
Forstwissenschaft*, **5**, 417–74.

Seeley, T. D. and Heinrich, B. (1981). Regulation of temperature in the nests of
social insects. In *Insect thermoregulation*, (ed. B. Heinrich), pp. 160–234.
Wiley-Interscience, New York.

Seppanen, E. J. (1969). Suurperhostemme talvehtimisasteet. *Annales
Entomologici Fennici*, **35**, 129–52.

Shands, W. A., Wave, H. E. and Simpson, G. W. (1961). Observations on egg
production by aphid pests of potato in Maine. *Annals of the Entomological
Society of America*, **54**, 376–8.

Shapashnikov, G. Kh. (1981). *Populations and species in aphids and the need for
a universal species concept*. Special Publication, Research Branch,
Agriculture Canada.

Sharrat, B. S. and Glen, D. M. (1988a). Orchard floor management utilizing
soil-applied coal dust for frost protection. Part I. Potential microclimate
modification on radiation frost nights. *Agriculture and Forest Meteorology*,
43, 71–82.

Sharrat, B. S. and Glen, D. M. (1988b). Orchard floor management utilising
soil-applied coal dust for frost protection. Part II. Seasonal microclimate
effect. *Agriculture and Forest Meteorology*, **43**, 147–54.

Shimada, K. (1988). Ice-nucleating activity in the alimentary canal of the
freeze-tolerant prepupae of *Trichiocampus populi* (Hymenoptera:
Tenthredinidae). *Journal of Insect Physiology*, **35**, 113–20.

Shorthouse, J. D., Zuchlinski, J. A. and Courtin, G. M. (1980). Influence of
snow cover on the overwinter of three species of gall forming *Diplolepis*
(Hymenoptera: Cynipidae). *Canadian Entomologist*, **112**, 225–9.

Simpson, R. G. and Welborn, C. E. (1975). Aggregations of alfalfa weevils,

Hypera postica, convergent ladybirds, *Hippodamia convergens*, and other insects. *Environmental Entomology*, **4**, 193–4.

Skanland, H. T. and Sømme, L. (1981). Seasonal variation in cold hardiness of the apple psyllid *Psylla mali* (Schmidt) in Norway. *Cryo-Letters*, **2**, 87–92.

Smith, S. M. and Brust, R. A. (1971). Photoperiodic control of the maintenance and termination of larval diapause in *Wyeomyia smithii* Coq. (Diptera: Culicidae) with notes on oögenesis in the adult female. *Canadian Journal of Zoology*, **49**, 1065–73.

Sømme, L. (1961). On the overwintering of houseflies (*Musca domestica* L.) and stable flies (*Stomoxys calcitrans* (L.)) in Norway. *Norsk Entomologisk Tidsskrift*, **11**, 191–223.

Sømme, L. (1964). Effects of glycerol on cold hardiness in insects. *Canadian Journal of Zoology*, **42**, 89–101.

Sømme, L. (1965a). Changes in sorbitol content and supercooling points in overwintering eggs of the European red mite *Panonychus ulmi* (Koch). *Canadian Journal of Zoology*, **43**, 881–4.

Sømme, L. (1965b). Further observations on glycerol and cold hardiness in insects. *Canadian Journal of Zoology*, **43**, 765–70.

Sømme, L. (1967). The effect of temperature and anoxia on haemolymph composition and supercooling in three overwintering insects. *Journal of Insect Physiology*, **13**, 805–14.

Sømme, L. (1969). Mannitol and glycerol in overwintering aphid eggs. *Norwegian Journal of Entomology*, **16**, 107–11.

Sømme, L. (1976). Cold-hardiness of winter-active Collembola. *Norwegian Journal of Entomology*, **23**, 149–53.

Sømme, L. (1978). Nucleating agents in the haemolymph of the third instar larva of *Eurosta solidaginis* (Fitch) (Diptera: Tephritidae). *Norwegian Journal of Entomology*, **25**, 187–8.

Sømme, L. (1982). Supercooling and winter survival in terrestrial arthropods. *Comparative Biochemistry and Physiology*, **73A**, 519–43.

Sømme, L. and Block, W. (1982). Cold tolerance of Collembola at Signy Island, Maritime Antarctic. *Oikos*, **38**, 168–76.

Sømme, L. and Conradi-Larsen, E. M. (1977). Cold-hardiness of collembolans and oribatid mites from windswept mountain ridges. *Oikos*, **29**, 118–26.

Sømme, L. and Ostbye, E. (1969). Cold-hardiness in some winter-active insects. *Norsk Entomologist Tidsskrift*, **16**, 45–8.

Sømme, L. and Zachariassen, K. E. (1981). Adaptations to low temperature in high altitude insects from Mount Kenya. *Ecological Entomology*, **6**, 199–204.

Sparrow, L. A. D. (1974). Observations on aphid populations on spring-sown cereals and their epidemiology in south-east Scotland. *Annals of Applied Biology*, **77**, 79–84.

Städler, E. (1970). Beitrag zur Venntris der Diapause bei der Möhrenfbiege (*Psila rosae* Fabr., Diptera: Psilidae) *Mitteilungen der Schweizerischen Entomologischen Gesellschaft*, **43**, 17–37.

Stark, R. W. (1959a). Climate in relation to winter mortality of the lodgepole needle miner, *Recurvaria starki* Free., in Canadian Rocky Mountain parks. *Canadian Journal of Zoology*, **37**, 753–61.

Stark, R. W. (1959b). Population dynamics of the lodgepole needle miner, *Recurvaria starki* Freeman, in Canadian Rocky Mountain parks. *Canadian Journal of Zoology*, **37**, 917–43.

Stevens, L. and McCauley, D. E. (1989). Mating prior to overwintering in the imported willow leaf beetle, *Plagiodera versicolora* (Coleoptera: Chrysomelidae). *Ecological Entomology*, **14**, 219–23.

Stoakley, J. T. (1977). A severe outbreak of the pine beauty moth on lodgepole pine in Sutherland. *Scottish Forestry*, **31**, 113–25.

Stoakley, J. T. (1979). *Pine beauty moth. Forest Record*, **120**, HMSO, London.

Stoakley, J. T. (1989). Forest insect pests and their control in relation to changes in land use and silvicultural practice. *Chemistry and Industry*, **20 March**, 186–90.

Storey, K. B. (1984). A metabolic approach to cold hardiness in animals. *Cryo-Letters*, **5**, 147–61.

Storey, K. B., Baust, J. G. and Buescher, P. (1981). Determination of water 'bound' by subcellular components during low temperature acclimation in the gall larvae, *Eurosta solidaginis*. *Cryobiology*, **18**, 315–21.

Strahler, A. N. (1963). *The Earth Sciences*. Harper and Row, New York.

Stroyan, H. L. G. (1977). Homoptera Aphidoidea Chaitophoridae and Callaphididae. *Handbook for the identification of British insects*, vol. 2 (4a). Royal Entomological Society, London.

Sullivan, C. R. (1965). Laboratory and field investigations on the ability of eggs of the European pine sawfly, *Neodiprion sertifer* (Geoffroy) to withstand low winter temperatures. *Canadian Entomologist*, **97**, 978–93.

Sutherland, O. R. W. (1968). Dormancy and lipid storage in the pemphigine aphid *Thecabius offinis*. *Entomologia Experimentalis et Applicata*, **11**, 348–54.

Suzuki, T. (1981). Effect of photoperiod in male egg production by foundresses of *Polistes chinennis antennalis* Perez (Hymenoptera, Vespidae). *Japanese Journal of Ecology*, **31**, 347–51.

Takeda, M. and Chippendale, G. M. (1982). Environmental and genetic control of the larval diapause of the Southwestern corn borer, *Diatraea grandiosella*. *Physiological Entomology*, **7**, 99–110.

Takehara, I. (1963). Glycerol in a slug caterpillar. II. Effect of some reagents on glycerol formation. *Low Temperature Science, Series B*, **21**, 55–60. (English summary).

Takehara, I. and Asahina, E. (1960). Glycerol in the overwintering prepupa of slug moth, a preliminary note. *Low Temperature Science, Series B*, **17**, 159–63. (English summary).

Takehara, I. and Asahina, E. (1961). Glycerol in a slug caterpillar. I. Glycerol formation, diapause and frost-resistance in insects reared at various graded temperatures. *Low Temperature Science, Series B*, **19**, 29–36.

Tamada, T. and Harrison, B. D. (1981). Quantitative studies on the uptake and retention of potato leaf-roll virus by aphids in laboratory and field conditions. *Bulletin of Entomological Research*, **77**, 135–44.

Tanaka, S. (1983). Seasonal control of nymphal diapause in the spring ground cricket *Pteromemobius nitidus* (Orthoptera: Gryllidae). In *Diapause and life cycle strategies in insects*, (ed. V. K. Brown and I. Hodek), pp. 33–53. Dr W. Junk Publishers, The Hague.

Tanaka, S. (1986). Uptake and loss of water in diapause and non-diapause eggs of crickets. *Physiological Entomology*, **11**, 343–51.

Tanno, K. (1964). High sugar levels in the solitary bee *Ceratina*. *Low Temperature Science, Series B*, **22**, 51–7.

Tauber, C. A. and Tauber, M. J. (1981). Insect seasonal cycles, genetics and evaluation. *Annual Review of Ecological Systematics*, **12**, 281–308.

Tauber, C. A. and Tauber, M. J. (1982). Evolution of seasonal adaptations and
    life history traits in *Chrysopa*: response to diverse selective pressures. In
    Evolution and genetics of life histories, (ed. H. Dingle and J. P. Hegmann),
    pp. 51–72. Springer-Verlag, New York.
Tauber, M. J. and Tauber, C. A. (1970). Photoperiodic induction and
    termination of diapause in an insect: response to changing daylengths.
    Science, **167**, 170.
Tauber, M. J. and Tauber, C. A. (1973a). Insect phenology: criteria for
    analysing dormancy and for forecasting postdiapause development and
    reproduction in the field. *Search (Agriculture)*, Cornell University
    Agriculture Experiment Station, Ithaca, New York, 3, 1–16.
Tauber, M. J. and Tauber, C. A. (1973b). Quantitative response to daylength
    during diapause in insects. *Nature*, **244**, 296–7.
Tauber, M. J. and Tauber, C. A. (1973c). Nutritional and photoperiodic control
    of the seasonal reproductive cycle in *Chrysopa mohave* (Neuroptera).
    Journal of Insect Physiology, **19**, 729–36.
Tauber, M. J. and Tauber, C. A. (1973d). Seasonal regulation of dormancy in
    Chrysopa carnea (Neuroptera). *Journal of Insect Physiology*, **19**, 1455–63.
Tauber, M. J. and Tauber, C. A. (1974). Thermal accumulations, diapause and
    oviposition in a conifer-inhabiting predator, *Chrysopa harrisii*
    (Neuroptera). *Canadian Entomologist*, **106**, 969–78.
Tauber, M. J. and Tauber, C. A. (1976a). Insect seasonality, diapause
    maintenance, termination and post-diapause development. *Annual Review
    of Entomology*, **21**, 81–107.
Tauber, M. J. and Tauber, C. A. (1976b). Environmental control of
    univoltinism and its evolution in an insect species. *Canadian Journal of
    Zoology*, **54**, 260–6.
Tauber, M. J., Tauber, C. A. and Denys, C. J. (1970). Adult diapause in
    Chrysopa carnea: photoperiodic control of duration and colour. *Journal of
    Insect Physiology*, **16**, 949–55.
Tauber, M. J., Tauber, C. A., Nechols, J. R. and Halgesen, R. G. (1982). A new
    role for temperature in insect dormancy: cold maintains diapause in
    temperate zone Diptera. *Science*, **218**, 690–1.
Tauber, M. J., Tauber, C. A., Nechols, J. R. and Obrycki, J. J. (1983). Seasonal
    activity of parasitoids: control by external, internal and genetic factors. In
    Diapause and life cycle strategies in insects, (ed. Brown, V. K. and Hodek,
    I.), pp. 87–108. Dr W. Junk Publishers, The Hague.
Tauber, M. J., Tauber, C. A. and Masaki, S. (1986). *Seasonal adaptations of
    insects*. Oxford University Press, New York.
Tauber, M. J., Tauber, C. A., Obrycki, J. J., Gallands, B. and Wright, R. J.
    (1988a). Voltinism and the induction of aestival diapause in the Colorado
    potato beetle, *Leptinotarsa decemlineata* (Coleoptera: Chrysomelidae).
    Annals of the Entomological Society of America, **81(5)**, 748–54.
Tauber, M. J., Tauber, C. A., Obrycki, J. J., Gallands, B. and Wright, R. J.
    (1988b). Geographical variation in responses to photoperiod and
    temperature by *Leptinotarsa decemlineata* (Coleoptera: Chrysomelidae)
    during and after dormancy. *Annals of the Entomological Society of America*,
    81, 764–73.
Taylor, F. and Spalding, J. B. (1988). Fitness functions for alternative
    developmental pathways in the timing of diapause induction. *American
    Naturalist*, **131**, 678–99.
Tenow, O. (1975). Topographical dependence of an outbreak of *Oporinia*

autumnata Bkh. (Lep., Geometridae) in a mountain birch forest in northern Sweden. *Zoon*, **3**, 85–110.

Tenow, O. and Nilssen, A. (1990). Egg cold-hardiness and topoclimatic limitations to the outbreaks of *Epirrita autumnata* in northern Fennoscandia. *Journal of Applied Ecology*, **27**, 723–734.

Tercafs, R. and Thines, G. (1973). Étude des dédencheurs visuels intervenant lors de la pénétration souterraine de *Scolopteryx libatrix* L. et *Triphosa dubitata* L. (Lepidoptères trogloxénes). *Annales de Speleologie*, **28**, 177–81.

Thomas, C. R. (1960). The European wasp (*Vespula germanica*) Fab. in New Zealand. *New Zealand Department of Science and Industry Research Information Service*, **27**, 1–27.

Topp, W. (1978). Untersuchungen zur Kalteresistenz bei staphyliniden (Col.) *Zoologischer Anzeiger*, **201**, 397–402.

Tsutsui, H., Hirai, Y., Honma, K., Tanno, K., Shimada, K. and Sakagami, S. (1988). Aspects of overwintering in the cabbage armyworm, *Mamestra brassicae* (Lepidoptera, Noctuidae) I. Supercooling points and contents of glycerol and trehalose in pupae. *Applied Entomology and Zoology*, **23**, 52–7.

Tsuji, H. (1966). Rice bran extracts effective in terminating diapause in *Plodia interpunctella* Hübner (Lepidoptera: Pyralidae). *Applied Entomology and Zoology*, **1**, 51.

Turl, L. A. D. (1980). An approach to forecasting the incidence of potato and cereal aphids in Scotland. *EPPO Bulletin*, **10**, 135–41.

Turl, L. A. D. (1983). The effect of winter weather on the survival of aphid populations on weeds in Scotland. *EPPO Bulletin*, **13**, 139–43.

Turl, L. A. D. (1987). A preliminary analysis of the relationship between aphid abundance and virus spread in Scottish seed potato crops. *Proceedings of Crop Protection in Northern Britain*, pp. 159–66.

Turnock, W. J. and Bilodeau, R. J. (1984). Survival of pupae of *Mamestra configurata* (Lepidoptera: Noctuidae) and two of its parasites in untilled and tilled soil. *Canadian Entomologist*, **116**, 257–67.

Turnock, W. J. and Philip, H. G. (1977). The outbreak of bertha armyworm *Mamestra configurata* (Noctuidae: Lepidoptera) in Alberta, 1971–1975. *Manitoba Entomologist*, **11**, 10–21.

Turnock, W. J., Lamb, R. J. and Bodnaryk, R. P. (1983). Effects of cold stress during pupal diapause on the survival and development of *Mamestra configurata* (Lepidoptera: Noctuidae). *Oecologia*, **56**, 185–92.

Umeozor, O. C., Van Duyn, J. W., Bradley, J. R. and Kennedy, G. G. (1985). Comparison of the effect of minimum-tillage treatments on the overwintering emergence of European corn borer (Lepidoptera: Pyralidae) in cornfields. *Journal of Economic Entomology*, **78**, 937–9.

Valder, S. M., Hopkins, T. L. and Valder, S. A. (1969). Diapause induction and changes in lipid composition in diapausing and reproducing face flies, *Musca autumnalis*. *Journal of Insect Physiology*, **15**, 1199–214.

Van den Berg, M. A. (1971). Studies on the induction and termination of diapause in *Mesocomys pulchriceps* Cam. (Hymenoptera: Eupelemidae) an egg parasite of Saturniidae (Lepidoptera). *Phytophylactica*, **3**, 85–8.

Van der Laak, S. (1982). Physiological adaptations to low temperature in freezing-tolerant *Phyllodecta laticollis* beetles. *Comparative Biochemistry and Physiology*, **73A**, 613–20.

Van der Woude, H. A. and Verhoef, H. A. (1986). A comparative study of winter survival in two temperate Collembola. *Ecological Entomology*, **11**, 333–40.

Van der Woude, H. A. and Verhoef, H. A. (1988). Reproductive diapause and cold hardiness in temperate Collembola *Orchesella cincta* and *Tomocerus minor*. *Journal of Insect Physiology*, **34**, 387–92.

Van Harten, A. (1983). The relation between aphid flight and the spread of potato virus Yn (PvYn) in the Netherlands. *Potato Research*, **26**, 1–15.

Van Houten, Y. M. (1989). Photoperiodic control of adult diapause in the predacious mite, *Amblyseius potentillae*: repeated diapause induction and termination. *Physiological Entomology*, **14**, 341–8.

Van Houten, Y. M., Overmeer, W. P. J. and Veerman, A. (1987). Thermoperiodically induced diapause in a mite in constant darkness is vitamin A dependent. *Experientia*, **43**(8), 933–5.

Van Kirk, J. R. and AliNiazee, M. T. (1981). Determining low-temperature threshold for pupal development of the western cherry fruit fly for use in phenology models. *Environmental Entomology*, **10**, 968–71.

Van Kirk, J. R. and AliNiazee, M. T. (1982). Diapause development in the western cherry fruit fly *Rhagoletis indifferens* Curran (Diptera, Tephritidae). *Zeitschrift für angewandte Zoologie*, **93**, 440–5.

Varley, G. C. and Butler, C. G. (1933). The acceleration of development of insects by parasitism. *Parasitology*, **25**, 263–8.

Vepsäläinen, K. (1974). The life cycles and wing lengths of Finnish *Gerris* Fabr. species (Heteroptera, Gerridae). *Acta Zoologica Fennica*, **141**, 1–73.

Waggoner, P. E. (1985). How gypsy moth eggs freeze. *Agricultural and Forest Meteorology*, **36**, 43–54.

Wall, C. (1974). Effect of temperature on embryonic development and diapause in *Chesias legatella*. *Journal of Zoology, London*, **172**, 147–68.

Walsh, P. J. (1990). *The influence of site factors on the community structures of pine beauty moth predators*. PhD Thesis, University of Ulster, Coleraine, Northern Ireland.

Walters, K. F. A. (1987). Forecasting the immigration of aphids into potato crops. Conference on Pests in Agriculture, Paris, pp. 587–94.

Walters, K. F. A. and Dewar, A. M. (1986). Overwintering strategy and the timing of the spring migration of the cereal aphids *Sitobion avenae* and *Sitobion fragariae*. *Journal of Applied Ecology*, **23**, 905–15.

Walters, K. F. A., Dixon, A. F. G. and Eagles, G. (1984). Non-feeding by adult gynoparae of *Rhopalosiphum padi* and its bearing on the limiting resource in the production of sexual females in host alternating aphids. *Entomologia Experimentalis et Applicata*, **36**, 9–12.

Wang, T. and Laing, J. E. (1989). Diapause termination and morphogenesis of *Holcotherax testaceipes* Ratzeburg (Hymenoptera: Encyrtidae), an introduced parasitoid of the spotted tentiform leafminer, *Phyllonorycter blancardella* (F.) (Lepidoptera: Gracillariidae). *Canadian Entomologist*, **121**, 65–74.

Ward, S. A., Leather, S. R. and Dixon, A. F. G. (1984). Temperature prediction and the timing of sex in aphids. *Oecologia*, **62**, 230–3.

Wasylyk, J. M., Tice, A. and Baust, J. G. (1988). Partial glass formation: a novel mechanism of cryoprotection. *Cryobiology*, **25**, 451–8.

Watabe, H. A. (1988). Photoperiod response of domestic *Drosophila funebris* (Diptera: Drosophilidae), with reference to its adaptive significance. *Kontyû*, **56**(3), 667–73.

Watt, A. D. (1987). Pine beauty moth outbreaks: the influence of host species, plant phenology, soil type, plant stress and natural enemies. In *Population biology and control of the pine beauty moth,* (ed. S. R. Leather, J. T. Stoakley and H. F. Evans), Forestry Commission Bulletin **67**, pp. 21–6. HMSO, London.

Watt, A. D. and Leather, S. R. (1988). The pine beauty moth in Scottish lodgepole pine plantations. In *Dynamics of forest insect populations: patterns, causes, implications,* (ed. A. A. Berryman), pp. 243–66. Plenum Press, New York.

Way, M. J. and Banks, C. J. (1964). Natural mortality of eggs of the black bean aphid. *Aphis fabae* on the spindle tree, *Euonymus europaeus. Annals of Applied Biology,* **54**, 255–67.

Way, M. J. and Banks, C. J. (1968). Population studies on the active stages of the black bean aphid, *Aphis fabae* Scop., on its winter host *Euonymus europaeus* L. *Annals of Applied Biology,* **62**, 177–97.

Way, M. J. and Cammell, M. E. (1982). The distribution and abundance of the spindle tree, *Euonymus europaeus,* in southern England, with particular reference to forecasting infestations of the black bean aphid, *Aphis fabae. Journal of Applied Ecology,* **19**, 929–40.

Way, M. J. and Heathcote, G. D. (1966). Interactions of crop density of field beans, abundance of *Aphis fabae* Scop., virus incidence and aphid control by chemicals. *Annals of Applied Biology,* **57**, 409–23.

Way, M. J., Cammell, M. E., Alford, D. V., Gould, H. J., Graham, C. W., Lane, A., Light, W. St. G., Rayner, J. M., Heathcote, G. D., Fletcher, K. E. and Seal, K. (1977). Use of forecasting in chemical control of black bean aphid, *Aphis fabae* Scop. on spring sown field beans, *Vicia faba* L. *Plant Pathology,* **26**, 1–7.

Way, M. J., Cammell, M. E., Taylor, L. R. and Woiwod, I. P. (1981). The use of egg counts and suction trap samples to forecast the infestation of spring sown field beans, *Vicia faba* by the black bean aphid, *Aphis fabae. Annals of Applied Biology,* **98**, 21–34.

Wellington, W. G. (1952). Air-mass climatology on Ontario north of Lake Huron and Lake Superior before outbreaks of the spruce budworm, *Choristoneura fumiferana* (Clem.) and the forest tent caterpillar, *Malacosoma disstria* Hbn. (Lepidoptera: Tortricidae; Lasiocampidae). *Canadian Journal of Zoology,* **30**, 114–27.

Wellington, W. G. (1954a). Weather and climate in forest entomology. *Meteorological Monographs,* **2**, 11–18.

Wellington, W. G. (1954b). Atmospheric circulation processes and insect ecology. *Canadian Entomologist,* **86**, 312–33.

Wellington, W. G. and Trimble, R. M. (1984). Weather. In *Ecological entomology,* (ed. C. B. Huffaker and R. L. Rabb), pp. 400–25. John Wiley & Sons Ltd, New York.

Wellington, W. G., Fettes, J. J., Turner, K. B. and Balyea, R. M. (1950). Physical and biological indicators of the development of outbreaks of the spruce budworm, *Choristoneura fumiferana* (Clem.) (Lepidoptera: Tortricidae). *Canadian Journal of Research,* **D28**, 308–31.

Werner, R. A. (1977). Biology and behaviour of the spear-marked black moth, *Rheumaptera hastata* in interior Alaska. *Annals of the Entomological Society of America,* **70**, 328–36.

Werner, R. A. (1978). Over-winter survival of spear-marked black moth, *Rheumaptera hastata* (Lepidoptera: Geometridae), pupae in interior Alaska. *Canadian Entomologist,* **110**, 877–82.

West, A. S. (1936). Winter mortality of larvae of the European pine shoot moth, *Rhyacionia buoliana* Schiff., in Connecticut. *Annals of the Entomological Society of America*, **29**, 438–48.

Wetzel, B. W., Kulman, H. M. and Witter, J. A. (1973). Effects of cold temperatures on hatching of the forest tent caterpillar, *Malacosoma disstria* (Lepidoptera: Lasiocampidae). *Canadian Entomologist*, **105**, 1145–9.

Whitfield, G. H., Drummond, F. A. and Haynes, D. L. (1986). Overwintering survival of the onion maggot, *Delia antiqua* (Meigen), (Diptera: Anthomyiidae). *Journal of the Kansas Entomological Society*, **59**, 197–8.

Wickman, B. E., Mason, R. R. and Trostle, G. C. (1981). *Douglas-fir tussock moth*. USDA Forest Service Forest Insect Disease Pest Leaflet **86**.

Wiklund, C. (1975). The evolutionary relationship between adult oviposition preferences and larval host plant range in *Papilio machaon* L. *Oecologia*, **18**, 185–97.

Wiktelius, S. (1982). Flight phenology of cereal aphids and possibilities of using suction trap catches as an aid in forecasting outbreaks. *Swedish Journal of Agricultural Research*, **12**, 9–16.

Wiktelius, S. (1988). Why is the bird cherry-oat aphid a pest in Sweden? *Ecological Bulletin*, **39**, 114–15.

Williams, C. T. (1980). Low temperature mortality of cereal aphids. *International Organisation for Biological Control Bulletin*, **1980/III/4**, 63–6.

Wilson, G. R. and Horsfall, W. R. (1970). Eggs of floodwater mosquitoes XII. Installment hatching of *Aedes vexans* (Diptera: Culicidae). *Annals of Entomological Society of America*, **63**, 1644–7.

Wipking, W. and Neumann, D. (1986). Polymorphism in the larval hibernation strategy of the burnet moth, *Zygaena trifolii*. In *Evolution of insect life cycles*, (ed. F. Taylor and R. Karban), pp. 125–34. Springer-Verlag, New York.

Witteveen, J., Verhoef, H. A. and Hupen, T. E. A. M. (1988). Life history strategy and egg diapause in the intertidal Collembolan *Anurida maritima*. *Ecological Entomology*, **13**, 443–51.

Woodford, J. A. T. and Lerman, P. M. (1974). Morphological variation in spring migrants of *Myzus persicae* (Sulz.) (Hemiptera, Aphididae): comparison of alatae from peach and marigold. *Bulletin of Entomological Research*, **64**, 595–604.

Worland, M. R. and Block, W. (1986). Survival and water loss in some Antarctic arthropods. *Journal of Insect Physiology*, **32**, 579–84.

Wright, D. W. and Ashby, D. G. (1946). Bionomics of the carrot fly (*Psila rosae* Fab.) II. Soil populations of carrot fly during autumn, winter and spring. *Annals of Applied Biology*, **33**, 263–76.

Wyatt, G. R. and Kalf, G. F. (1957). The chemistry of insect haemolymph II: Trehalose and other carbohydrates. *Journal of General Physiology*, **40**, 833–47.

Wyatt, G. R. and Kalf, G. F. (1958). Organic components of insect haemolymph. *Proceedings X International Congress of Entomology* (*Montreal*) *1956, vol. 2*, p. 333.

Wyatt, G. R. and Meyer, W. L. (1959). The chemistry of insect haemolymph III. Glycerol. *Journal of General Physiology*, **42**, 1005–11.

Wylie, H. G. (1977). Preventing and terminating pupal diapause in *Alkrycia cinerea* (Diptera: Tachinidae). *Canadian Entomologist*, **109**, 1083–90.

Wylie, H. G. (1980). Factors affecting facultative diapause of *Microctonus vittatae* (Hymenoptera: Braconidae). *Canadian Entomologist*, **112**, 747–9.

Yata, O. (1974). Studies on seasonal forms and imaginal diapause in two

Eurema species in Japan (Lepidoptera: Pieridae) (English summary). *Tgyo to Ga*, **25**, 47–54.

Young, S. R. and Block, W. (1980). Experimental studies on the cold tolerance of *Alaskozetes antarcticus*. *Journal of Insect Physiology*, **26**, 189–200.

Zachariassen, K. E. (1977). Effects of glycerol in freeze-tolerant *Pytho depressus* L. (Col., Pythidae). *Norwegian Journal of Entomology*, **24**, 25–9.

Zachariassen, K. E. (1979). The mechanism of the cyroprotective effect of glycerol in beetles tolerant to freezing. *Journal of Insect Physiology*, **25**, 29–32.

Zachariassen, K. E. (1980). The role of polyols and nucleating agents in cold-hardy beetles. *Journal of Comparative Physiology*, **140**, 227–34.

Zachariassen, K. E. (1982). Nucleating agents in cold-hardy insects. *Comparative Biochemistry and Physiology*, **73A**, 557–62.

Zachariassen, K. E. (1985). Physiology of cold tolerance in insects. *Physiological Reviews*, **65**, 799–832.

Zachariassen, K. E. and Hammel, H. T. (1976). Nucleating agents in the haemolymph of insects tolerant to freezing. *Nature*, **262**, 285–7.

Zachariassen, K. E. and Husby, J. A. (1982). Antifreeze effect of thermal hysteresis agents protects highly supercooled insects. *Nature*, **298**, 865–7.

Zachariassen, K. E. and Pasche, A. (1976). Effect of anaerobiosis on the adult cerambycid beetle *Rhagium inquisitor* L. *Journal of Insect Physiology*, **22**, 1365–8.

Zachariassen, K. E., Hammel, H. T. and Schmidek, W. (1979). Studies on freezing injuries in *Eleodes blanchardi* beetles. *Comparative Biochemistry and Physiology*, **63A**, 199–202.

Zaslavsky, V. A. (1984). *Photoperiodic and temperature control in insect development*. Nanka, Leningrad.

Zaslavsky, V. A. (1988). *Insect development – photoperiodic and temperature control*. Springer-Verlag, Berlin, Heidelberg.

Zaslavsky, V. A. and Umarova, R. Ya. (1981). Photoperiodic and temperature control of diapause in *Trichogramma evanescens* West. (Hymenoptera, Trichogrammatidae). *Entomological Review*, **60**, 1–12.

Ziegler, R. and Wyatt, G. R. (1975). Phosphorylase and glycerol production activated by cold in diapausing silk moth pupae. *Nature*, **254**, 622–3.

Index

Printed in the United States
By Bookmasters